Energy Autonomy of Batteryless and
Wireless Embedded Systems

Energy Management in Embedded Systems Set

coordinated by
Maryline Chetto

Energy Autonomy of Batteryless and Wireless Embedded Systems

Aeronautical Applications

Jean-Marie Dilhac
Vincent Boitier

First published 2016 in Great Britain and the United States by ISTE Press Ltd and Elsevier Ltd

ISTE Press Ltd
27-37 St George's Road
London SW19 4EU
UK

www.iste.co.uk

Elsevier Ltd
The Boulevard, Langford Lane
Kidlington, Oxford, OX5 1GB
UK

www.elsevier.com

Notices
Knowledge and best practice in this field are constantly changing. As new research and experience broaden our understanding, changes in research methods, professional practices, or medical treatment may become necessary.

Practitioners and researchers must always rely on their own experience and knowledge in evaluating and using any information, methods, compounds, or experiments described herein. In using such information or methods they should be mindful of their own safety and the safety of others, including parties for whom they have a professional responsibility.

To the fullest extent of the law, neither the Publisher nor the authors, contributors, or editors, assume any liability for any injury and/or damage to persons or property as a matter of products liability, negligence or otherwise, or from any use or operation of any methods, products, instructions, or ideas contained in the material herein.

For information on all our publications visit our website at http://store.elsevier.com/

British Library Cataloguing-in-Publication Data
A CIP record for this book is available from the British Library
Library of Congress Cataloging in Publication Data
A catalog record for this book is available from the Library of Congress
ISBN 978-1-78548-123-9

Printed and bound in the UK and US

Contents

Foreword 1

At a time when digital transformation is at the heart of significant changes to the majority of traditional industrial sectors, it is good to return to the principles of elementary physics and its fundamental laws. This volume by Jean-Marie Dilhac and Vincent Boitier reminds us that before becoming virtual, all information came from the combination of known, concrete, physical phenomena, which systems refer to as *sensors* in this volume, making it possible to retrieve information in real time and to transmit to other systems capable of analyzing this information.

Sensors thus make it possible to make objects, materials and structures which had previously been thought of as passive elements "speak". The exponential development of the Internet of Things and the proliferation of these passive objects which are now connected has led to a range of applications made possible by these sensors. Logistics, aeronautical maintenance, self-driving cars, control of aerostructures, health ... these are just a few examples where the use of sensors is the basis for a new range of services, solutions and economic models.

However, there are two major constraints which existing systems come up against: energy autonomy and the means of transmitting information once it has been obtained. Furthermore, for the aeronautical industry, there is a third constraint: weight.

The work presented here provides a situational analysis of the physical laws which it is possible to make use of, in order to develop autonomous sensors (not connected to an electric network or to a battery). Depending on how they are to be used and their respective efficiency, potential sensors can have various characteristics. However, the range of techniques used demonstrates that all kinds of application are possible.

Among the models of energy autonomy that are brought up, thermogenerators have been identified as having a potential use in aeronautics. The authors were among the first involved in aeronautical developments in the retrieval of energy through thermoelectricity with the Airbus Group. The ongoing research of examples of applications and industrial uses is one of the strengths of this volume. Indeed, on top of the assessment of technical possibilities, the authors aim to inform the reader of the contexts for putting various types of autonomous sensors into practical use. Manufacturers, especially in the field of aeronautics – engine designers, manufacturers, interior designers – will thus find avenues of exploration for their design specifications. In this way, examples of use during test flights will give a rough sketch of the industrial applications of this new range of energy self-sufficient sensors.

The stakes are high:

– first, ubiquity. Autonomous sensors do not need to be accessible in order to be maintained or recharged. Thus, they can be placed in the most sensitive areas, often out of reach. In terms of the monitoring of the structure, this enables a better level of sensor coverage;

– second, there is the question of temporal permanence. Autonomous sensors are constantly active. This makes it possible to imagine new scenarios for applications where the integration of these sensors will be carried out at the point where the program (aircraft or otherwise) is conceived, because the sensors will have a lifespan which will be close to that of the platform itself;

– third, there are safety challenges. More sensors make it possible to monitor systems more effectively and thus to better prevent risks. Platforms will then increase their own capacity to signal a fault;

– finally, we have the new economic models. Once it is possible to integrate self-sufficient sensors on platforms on an industrial level, a reduction in maintenance costs will be the first advantage. Depending on the industry concerned, it will be possible to develop new models based on parameters monitored and transmitted by these sensors.

In order to increase the range of applications for energy autonomous sensors in the field of aeronautics, it will be necessary to continue research with a view to reducing the weight factor for systems for producing energy using the ambient environment. Indeed, these should be able to contribute "product/mass energy" outputs that will allow them to be taken into account for affordable operating costs. This will doubtless be the subject of future study.

Frédéric SUTTER
"Digital Transformation" Program Director, Airbus Group
August 2016

Foreword 2

Our societies are going through a genuine digital revolution, known as the Internet of Things (IoT). This is characterized by increased connectivity between all kinds of electronic materials, from surveillance cameras to implanted medical devices. With the coming together of the worlds of computing and communication, embedded computing now stretches across all sectors, public and industrial. The IoT has begun to transform our daily life and our professional environment by enabling better medical care, better security for property and better company productivity. However, it will inevitably lead to environmental upheaval, and this is something that researchers and technologists will have to take into account in order to devise new materials and processes.

Thus, with the rise in the number of connected devices and sensors, and the increasingly large amounts of data being processed and transferred, demand for energy will also increase. However, global warming is creating an enormous amount of pressure on organizations to adopt strategies and techniques at all levels that prioritize the protection of our environment and look to find the optimal way of using the energy available on our planet.

Over the past few years, the main challenge facing R&D has been what we have now come to refer to as green electronics/computing: in other words, the need to promote technological solutions that are energy efficient and that respect the environment.

The set of books entitled *Energy Management in Embedded Systems* has been written in order to address this concern:

In their volume *Energy Autonomy of Batteryless and Wireless Embedded Systems,* Jean-Marie Dilhac and Vincent Boitier consider the question of the energy autonomy of embedded electronic systems, where the classical solution of the electrochemical storage of energy is replaced by the harvesting of ambient energy. Without limiting the comprehensiveness of their work, the authors draw on their experience in the world of aeronautics in order to illustrate the concepts explored.

The volume *ESD Protection Methodologies,* by Marise Bafleur, Fabrice Caignet and Nicolas Nolhier, puts forward a synthesis of approaches for the protection of electronic systems in relation to electronic discharges (ElectroStatic Discharge or ESD), which is one of the biggest issues with the durability and reliability of new technology. Illustrated by real case studies, the protection methodologies described highlight the benefit of a global approach, from the individual components to the system itself. The tools that are crucial for developing protective structures, including the specific techniques for electrical characterization and detecting faults as well as predictive simulation models, are also featured.

Maryline Chetto and Audrey Queudet present a volume entitled *Energy Autonomy of Real-Time Systems.* This deals with the small, real-time, wireless sensor systems capable of taking their energy from the surrounding environment. Firstly, the volume presents a summary of the fundamentals of real-time computing. It introduces the reader to the specifics of so-called *autonomous* systems that must be able to dynamically adapt their energy consumption to avoid shortages, while respecting their individual time restrictions. Real-time sequencing, which is vital in this particular context, is also described.

The volume entitled *Flash Memory Intrgration* by Jalil Boukhobza and Pierre Olivier attempts to highlight what is currently the most commonly used storage technology in the field of embedded systems. It features a description of how this technology is integrated into current systems and how it acts from the point of view of performance and energy consumption. The authors also examine how the energy consumption and the performance of a system are characterized at the software level (applications, operating system) as well as the material level (flash memory, main memory and CPU).

Maryline CHETTO

August 2016

Preface

For the past decade or so, we have been fortunate enough to work in the domain of wireless and therefore energy autonomous sensor networks, in a variety of fields such as agriculture, space exploration and aeronautics. We have thus focused more specifically on developing methods and techniques for controlling energy, the majority of the time for systems requiring the harvesting of ambient energy.

Rather quickly, our application scope become focused on aeronautics, and we had the opportunity to work on needs expressed by professionals working in the field. Some of these studies made it necessary to carry out specific measurements during test flights and the most recent of these led to flight test missions of complete systems.

It seemed appropriate, therefore, at a point where the study of energy autonomy in embedded sensor systems is appearing on university courses, to present a summary of our most significant conclusions in a volume which will allow students, engineers and researchers to understand the field of energy autonomous sensors, a field which is expanding but which remains marginal or even absent in several industrial sectors. The description of practical experiments, in the fourth chapter of this volume, is preceded by three introductory chapters, which will make it possible to place these examples in the wider context of wireless sensor networks, energy self-sufficiency and power electronics.

This volume should be seen as accessible for students, engineers and researchers, regardless of whether or not they are experts in electronics, materials physics, heat transfer or aeronautics. Conversely, given the diversity of the design specifications and the different technologies implemented, it is not able to cover the full range of directly applicable systems. Its aim is to enable the reader to advance further in their studies, and to help open the right doors for them.

Before concluding this preface, we must of course mention that the contents of this volume are not solely down to the work carried out by the two authors, but the result of meetings, collaborations and teamwork. In particular, we would like to thank the staff at Airbus who trusted us, our work colleagues, our doctoral students and our interns. There are so many of them that in making a list of names we would regretfully end up omitting someone. However, we hope they know that we are aware of their contribution and how thankful we are.

Jean-Marie DILHAC
Vincent BOITIER
August 2016

Wireless Sensor Networks

1.1. Brief historical perspective

Although it is currently the subject of much discussion, in research laboratories as well as in the industrial world, the concept of *wireless sensor networks* is quite old. We will not pretend that it was the first instance of its implementation, but it is interesting to note, for example, the deployment from 1967 onward of such sensors by the American army during the Vietnam War[1]. This technology continues to be used in a military context, with miniaturized systems which fall into the bracket of Unattended Ground Sensors (UGS).

A second noteworthy historical landmark, which led to the creation of another range of devices, was the publicity surrounding the new concepts developed by the University of California, Berkeley in 2003. Under the title *Smart Dust*, a structure for new systems was unveiled which, at the level of a cubic millimeter, drew together sensors, signal processing and wireless telecommunication (radio or optical). This new concept emerged from a number of studies undertaken and financed in the USA, in the previous decade, by the Rand Corporation (an American think tank) and the Defense Advanced Research Projects Agency (DARPA). This revolutionary structure benefits on one hand from the progress made in microelectronics and in

---

1 20,000 systems, equipped with batteries, were released by plane or by helicopter. Designed for detecting the passage of convoys, they were equipped with seismic or acoustic sensors. On landing, the bulk of the system sunk into the Earth, with only the radio antenna remaining above the surface. Planes relayed measurements to a control center.

miniaturization, and on the other hand from microelectromechanical systems (MEMS) technology, itself a more recent form of technology given that it originated in microelectronics. The main contribution of MEMS was how it could be used for miniaturized sensors or actuators. However, from the moment these studies were published, the Berkeley researchers admitted that it seemed as though questions relating on one hand to energy and how it could be stored and the wireless transmission of information[2] on the other hand, would present obstacles to the total miniaturization of the system.

The applications considered then concerned the spread of a wide range of these systems in clothing, building, civil engineering infrastructure, animals and mankind itself. However, this form of generalized invasion was and continues to be held up by two factors: limitations in terms of technology and perhaps most importantly the rarity of feasible applications which require the deployment of these ultra-miniaturized wireless sensor networks. As has already been stated, despite its issues the concept was highly stimulating for university research and industrial development laboratories [BEL 12] and today realistic applications exist, including the measurement of tire pressure using so-called Tire Pressure Measurement Systems (TPMS) systems or the remote monitoring of industrial machinery. However, the systems deployed operate somewhere in the region of a few cubic centimeters, without taking radio antennae into account.

1.2. Some principles and definitions

In this section, we will provide some references relating to the structure of a wireless network, and the functional structure of its nodes.

1.2.1. *The structure of a wireless sensor network*

A wireless sensor network draws together systems (unfairly termed *sensors*) which comprise the nodes of this network. There can be a large

---

2 Aside from the marginal example of active (the system has a modulated light source) or passive (a micro mirror reflects or does not reflect an incident light beam) optical transmission, radio transmission requires an antenna whose size is not properly compatible with the millimetric or micrometric scale.

number of these nodes or a relatively small number. Each node must be able to transmit information, gathered by its sensor(s), to a collection point. The transmission can be direct or can move around, and in this case each node (or a few) plays the role of a relay and eventually a router. In the case of a radio link, transmission *via* relay makes it possible to reduce the power radiated by the antenna, given that the power received by the antenna is a function of the inverse square of the distance between nodes[3]. Although this is tempting, the use of a relay is nevertheless a complicated method in terms of synchronizing the sending of frames, the management of collisions and the routing of messages.

Finally, some nodes can be used for aggregating data, in order to reduce transmission flows, and to avoid overcharging the relay nodes located at the vicinity of the collection point. Nodes will therefore not necessarily be identical.

The network can be deployed manually, where each node occupies a well-defined position, or by chance (natural dispersion). The topology of the network can be fixed (deployment in a building, a boat, an aircraft or a factory) or can evolve over time (deployment over several mobile objects, in a liquid environment, on cooperative drones). In situations where the nodes are considered inaccessible once they have been deployed (dispersed into nature, lost in a concrete structure, placed in an inaccessible area – aside from heavy disassembly – of an aircraft), it will be necessary to consider tolerance of errors and self-organization.

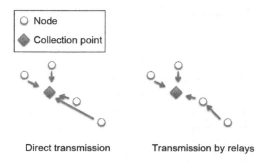

Direct transmission   Transmission by relays

Figure 1.1. *Principle structure of a wireless sensor network*

---

3 In free space. In a complex environment, with a number of obstacles, the weakening is even quicker, and increases the advantage of using a relay.

1.2.2. *Structure of a node*

A node of a wireless sensor network will have at the very least:

– a sensor;

– a unit for analog (processing the sensor's output, analog/digital conversion and radiofrequency levels) and digital processing (filtering, memorizing, clock and modulation) signals;

– a wireless transmission device;

– an energy source (a necessity given that the node is isolated from any cable network).

It might also be equipped with:

– a wireless reception device;

– an actuator;

– a specialized positioning unit (like GPS, which makes it possible to retrieve a temporal reference).

All this is in packaging which from the point of view of the design brief can be restricted by:

– the weight;

– the volume;

– environmental conditions (temperature, pressure, vibrations and accelerations, electromagnetic radiation, chemical substances, explosive environment, etc.).

The functioning of the unit is controlled by an operating system which, the majority of the time, must remain energy efficient while maintaining the desired functionality. We will explore this point in greater detail further on, but for those energy autonomous nodes, this is a resource which it is required to save. It is certainly necessary to install specific embedded software as well as adapted radio modulations. From a material point of view, one traditional approach is to only supply those parts of the system which are required in a given operating phase; for example, energy will not be supplied to radio circuits[4] when there is no expected transfer of data. However, special care must be taken when deploying this traditional strategy; it is a common occurrence for an integrated circuit to consume more energy during transient states of start-up or stopping than in steady state. It is therefore necessary to have an assessment which takes into account energy consumption in transient and steady states as well as the frequency of being switched on and off.

1.3. The energy question

The choice to use a wireless network means that each node will be energy autonomous, given that these nodes will not have access to an energy supply *via* a cable. An obvious solution is therefore to extract this energy from a reserve located in the node, a reserve which uses an electrochemical battery. A slightly more precise analysis will require other solutions, outlined in Figure 1.2 and we will now summarize these solutions from left to right (see Figure 1.2).

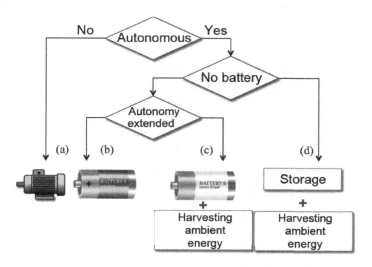

Figure 1.2. *Different options for supplying a node*

4 Generally speaking, these are the components which use up the most energy.

We will see further on that energy autonomy is a powerful constraint; it is therefore only natural to properly weigh the consequences of withdrawing the option of a wired network (option (a) in Figure 1.2) if only for its energy contribution. There are situations where an energy network, adapted to the needs of the node, will not be too far from it: for example, in the body of a satellite, or in parts of a land vehicle. In this type of context, energy autonomy can be more of a handicap (in terms of reliability, complexity or weight[5]) than an advantage.

However, in this volume, we will concern ourselves with wireless networks, and therefore the need for energy autonomy. The absence of wiring has several obvious advantages, including:

– much greater speed of deployment;

– where necessary, it is easier to modify the positioning of the node;

– in some cases (long cables), there will be a weight advantage as well as a reduction in costs due to the absence of wiring;

– lack of pollution (aside from electromagnetic pollution) of the energy network eventually present in the environment around the node (onboard network).

As has already been stated, the simplest solution is therefore to use an energy reserve provided by a non-rechargeable (primary) electrochemical battery (option (b) in Figure 1.2). Electrochemical storage is a well-developed technology and efficient lithium-based products are available which feature:

– high energy density[6] (roughly 500 Wh/kg);

– weak self-discharge (around 1% over a year);

– a lifespan of several years.

---

5 As part of their development activities, on several occasions, the authors of this volume found themselves asked to design a source of harvested ambient energy which did not weigh more than the cable which would connect an energy network situated a few centimeters or meters from the node.

6 Compared to the progress made in terms of integration density in electronic circuits (Moore's law), progress has been much slower in terms of energy density stored in a battery.

However, there are usage restrictions which should not be ignored:

– performance is highly dependent on temperature and the operating range is restricted (–60°C to + 85°C for an efficient battery, for example, although there is a strong increase in series resistance at low temperatures);

– energy density compared to that of chemical explosives, there is a great risk in case of fire or explosion[7] if standard usage guidelines (temperature and current) are not followed;

– finally and perhaps most importantly, the energy reserve is obviously limited, and the battery will need to be changed once its reserves have run out. This need to constantly change as well as limited autonomy can present a major obstacle depending on the way in which the sensor network is being used.

In order to increase the level of autonomy, or even to obtain full autonomy, it is possible to employ *ambient energy harvesting*. This principle is based on the transformation of a flow of energy naturally present in the environment around the node into electrical energy, through the intermediary of a specific transducer. Harvesting light energy through a photovoltaic procedure is a good illustration of this principle; as we will see, other sources of energy are also suitable. The way in which this energy transformation is achieved can take place in three different ways:

– the transducer operates in tandem with a non-rechargeable (primary) battery (option (c) in Figure 1.2) providing part of the energy consumed by the node and expands the system's autonomy;

– the transducer operates in tandem with a rechargeable (secondary) battery (also represented by option (c) in Figure 1.2), when it is possible to supply a sufficient amount of energy to the charge, with the rechargeable battery playing the role of intermediary (or buffer) reservoir;

– the transducer operates in tandem with an electrostatic storage system, with no requirement for chemical reactions (option (d) in Figure 1.2). We will examine this structure in greater detail further on, but for the time being we will note that while the storage of electrostatic energy by capacitor or supercapacitor gives a reduced level of performance in terms of energy

7 As a result, these high-performance batteries are equipped with passive protections (fuse, weak point connected to a vent allowing gas escaping in order to prevent an uncontrolled explosion).

density compared to electrochemical storage, it is less dangerous, and furthermore it enables a practically unlimited number of charge/discharge cycles.

It should also be noted that, for general use, there are more restrictions with rechargeable electrochemical batteries than non-rechargeable batteries: very often, it is not possible to recharge them at negative temperatures (below 0°C). Maintaining their lifespan also requires complex electronic circuits in order to control the charge (charge with constant current and/or voltage, detection of the end of charging, possibility of injecting or not injecting a topping current in order to compensate for the self-discharge). Finally, their energy density is weaker than non-rechargeable batteries, but there are the same risks if not properly used[8].

1.4. Aeronautics

Since the dawn of aeronautics, there have been a number of technological breakthroughs. The most recent concern the use of composite materials in structural parts and the increasingly important position of computerized, electronic and electrical systems. Civil aviation – despite the "crisis" – is a dynamic domain in a period of significant growth, both in terms of manufacturers and airline companies and airports, and can even be considered as one of the key contributors to the economy.

However, at the beginning of the 21st Century, aeronautics is faced with a number of challenges, including the increase in traffic, safety and security, environmental preservation (air pollution, noise and aircraft recycling), the energy crisis in the short (rising fuel prices) and long term (the end of fossil fuels[9]). In order to overcome these challenges, actions are currently being undertaken or planned relating to:

– air Traffic Management;

– airport design;

– engines (for example, *Open Rotor*);

8 More information on these aspects can be found in the manufacturers' catalogs (Saft: http://www.saftbatteries.com).
9 Such a gap exists in terms of energy density between kerosene and its substitutes (including electrochemical substances for electric drive) that the joke is often made that the last drop of petrol on Earth will be burned in an aircraft (certainly military, and probably American!).

– aerodynamics;

– construction materials, reduction of weight;

– more Electrical Aircraft (MEA).

The more electrical aircraft (MEA) is a concept in which traditional power networks located on an aircraft (mechanical, pneumatic and hydraulic), all supplied with energy by kerosene burned in the engines, will be gradually replaced by a power network that is largely electric (aside from propulsion, with this particular aspect relating to aircraft said to be *electric*). The aims pursued relate to the reduction of weight as well as reducing the cost of production and maintenance. The MEA concept is part of a wider framework that can be characterized by three stages:

– *Fly-by-Wire* (electric flight commands, introduced to civil aviation with Concorde and the Airbus A320);

– *Drive-by-Wire* (eliminating the physical, mechanical, pneumatic or hydraulic links between sensors and actuators);

– *Power by Wire* (electric actuators).

However, an electric actuator does not possess the latitude to evacuate by *fluid* any unwanted heat that has been generated locally (particularly in cases of high containment), as a hydraulic actuator is capable of doing. When the actuator is positioned in a wing that has been carefully designed to be air-tight in order to improve its aerodynamic performance, thus eliminating any air flow, it can be necessary to incorporate air ducts in order to cool down the electric actuator (Figure 1.3). The reader should not be surprised by these details which might seem slightly off-topic: they demonstrate that, provided the justification is excellent, an aircraft manufacturer can agree to locally disrupt the aerodynamic efficiency of its aircrafts. We will return to this point further on, in the chapter of this volume dealing with the use of aeroacoustics.

Generally speaking, within the wider context of innovation, sensor networks will play an increasingly important role: environmental control of the passenger cabin and connecting personal electronic equipment, helping pilots with decision-making, identifying the cargo and any detachable equipment (life preservers) through improved RFID tags, safe and optimal integration of the aircraft in its environment on the ground and in-flight.

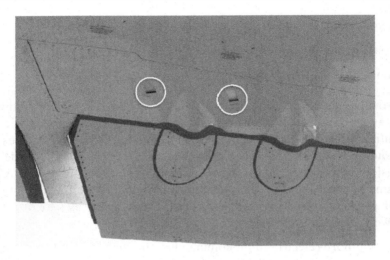

Figure 1.3. *Air ducts (in the two white circles) arranged
to allow the cooling of two aileron actuators when in electric
mode onboard an Airbus A350. The wing has been initially
carefully air-tighted (Paris Air Show 2015, authors' photograph)*

Sensor networks will also be used for monitoring the aging of the structure of the aircraft as well as its components. Essentially, an aircraft often finds itself in a harsh environment (variations in temperature and pressure, humidity, vibrations, dust, collisions, etc.), while for some of its components there are added specific constraints (acoustic charge and temperature in the vicinity of the engine and the brakes, mechanical constraints, corrosive liquids, etc.) It is therefore essential to carry out regular inspections and maintenance. Traditionally, these are carried out periodically, looking to anticipate predictable faults or decreased efficiency. Health Monitoring by sensor networks will make it possible to only carry out repairs when necessary, thus saving time (especially for areas which are difficult to access), weight (by decreasing mechanical margins) and therefore fuel.

However, the deployment of new wired sensor networks comes up against the complexity of electric networks already present in commercial aircraft. Figure 1.4 gives a few approximate values of cable length for a jetliner, a fighter jet and a modern car. In order to more precisely evaluate the importance of wiring, we have also included a comparison of the total length of cable per linear meter.

Figure 1.4. *Top scale: total length of electrical wiring in a modern car, a fighter jet and a jetliner. Bottom scale: total length in relation to the length of the given object (3.8, 16 and 73 m, respectively)*

From this figure, it is clear that the length of wiring in aeronautics is very large and that this involves a vast complexity of networks, which can be evaluated using the length of wiring compared to the linear meter of a plane or a car. Finally, in the case of the A380 where we are aware of this approximate value, it could be useful to mention that the additional length of cable used during test flights carried out by the manufacturer is 300 km, added to the 500 km already present. Within the framework of test flights, these additional networks are particularly difficult to deploy in an aircraft which has not been designed for this purpose. This difficulty will only increase for aircraft in which composite materials are increasingly used, due to the difficulty (or even impossibility) to create passages for cables in an opportune manner, for bulkheads or parts of the structure.

There is therefore a paradoxical element to aeronautics: the functional need of new sensor networks has increased, but comes up against the issue of not overcharging wiring which is already complex. It is therefore natural that wireless sensor networks should be required, provided that they are energetically autonomous, an aspect that we will consider in more detail in the following chapter.

2

Energy Autonomy

2.1. Introduction

In this chapter, we will outline the principles implemented in order to obtain energy autonomy for an isolated system. To begin with, we will compare electrochemical and electrostatic energy storage, forms of technology which are at the heart of energy autonomy for wireless sensor networks. In section 2.3, we will look at the architecture of systems based around harvesting ambient energy, and in section 2.4 we will consider the physical principles for the transducers used for energy capture, without forgetting the relevance of these systems in an aeronautical context. For the most part, the following chapter will focus on managing energy in electronic circuits. It should be noted that, for this current volume, we have limited ourselves to one application scope (aeronautics) within which we have chosen those structures which would appear to be the most relevant. For further information, we would direct the reader toward studies found in [BEL 12] and [SPIE 15].

2.2. Electrochemical source and electrostatic energy storage

In Chapter 1, we were able to identify three methods for attaining energy autonomy for one node of a wireless sensor network (Figure 1.2):

– the use of a non-rechargeable electrochemical battery (*primary battery*);

– the use of a rechargeable electrochemical battery (*secondary battery*) assisted by a system for harvesting ambient energy;

– harvesting ambient energy on its own, assisted by electrostatic energy storage (storage is always necessary for reasons outlined below).

In these three situations, it is necessary to use a form of energy storage, either electrochemical or electrostatic. We will elaborate on these three methods alongside those already discussed in the first chapter.

The electrochemical storage of electrical energy[1] was introduced in 1800 with the battery invented by Volta, an invention preceded by the work of the biologist Nicholson and the fortunate discovery of *Galvanism* by the physician Galvani. In 1859, Gaston Planté invented the first *rechargeable* lead-acid battery. Progress continued to be made in terms of ease of use and energy density up until the modern day with the introduction from 1990 onward of several kinds of battery technology, rechargeable or non-rechargeable using lithium ions[2].

Here, it will be useful to look in more detail at the often changing vocabulary of electrochemical energy storage:

– a galvanic cell is made up of two electrodes in an electrolyte bath; for the most part, these are not rechargeable (primary);

– an accumulator is conversely a rechargeable (secondary) device;

– a battery is an association of either two or more galvanic cells or accumulators.

In a more quantitative manner, a system for electrochemical energy storage can be characterized by a number of parameters, including the following:

– open-circuit voltage (V) which differs from the load voltage because of the *Equivalent Series Resistance* (Ω) (ESR), resistance on which the temperature has a significant impact (variation in the order of tens on the normal range of applications in temperature – within this range, ESR decreases with temperature);

– capacity (Ah);

---

1 We are aware that electricity cannot be stored directly, except in very small quantities in electrostatic or electrodynamic form.
2 Unsuccessful attempts were made beginning in 1970, but given that metallic lithium spontaneously combusts when exposed to air, it was necessary to master the use of lithium before producing commercial products. Lithium is therefore used in ion form, with inclusion in an insertion compound or with a polymer electrolyte.

– mass energy density (J/kg, Wh/kg) or volumetric energy density (J/l, Wh/l);

– mass power density (W/kg), or volumetric power density (W/l);

– self-discharge (expressed as a percentage over a given period);

– residual energy (energy which cannot be extracted);

– environmental conditions for storage and use, maximum currents, etc.

Energy and power densities are essential selection criteria. In order to facilitate comparisons between different storage methods and to differentiate between the various commercial products, it is regularly necessary to use a *Ragone chart*[3]. This is a graphic with logarithmic axes which gives the energy density as a function of the power density (or occasionally the inverse). One such diagram is used in Figure 2.1 for which, by way of illustration, we have chosen to include several storage components[4]. However, we have chosen not to include so-called *thermal*[5] batteries, the use of which as emergency energy sources in aircrafts is currently being studied.

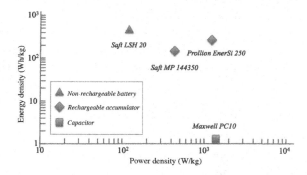

Figure 2.1. *Ragone chart giving the performances of batteries using lithium ions on one hand and an ultra-capacitor on the other hand*[6]

---

3 Named after David Ragone who proposed this representation in a scientific journal in 1968.
4 These are high-quality components, which do not represent the average performance in their respective categories.
5 These are electrochemical batteries which are initially inert, where the solid electrolyte passes to a liquid state due to the effect of a pyrotechnic element.
6 The data in this graphic, as well as the data used in the text, either came directly from, or were calculated using, data from manufacturers sometimes in extreme conditions. For instance, the power density of the LSH 20 battery can only be guaranteed for 100 ms.

We will start by illustrating our intentions with a high-performance non-rechargeable battery, the LSH 20 model from Saft[7]. Aside from the power and energy densities outlined in Figure 2.1, this component possesses a range of application of between −60 and +85°C, an annual self-discharge of 3% and an ESR of a few ohms with a weak current and at room temperature.

We will also look at a rechargeable accumulator, the MP 144350, which was also produced by SAFT[8]. During discharge, its range of functioning extends from −50 to +60°C, but is restricted to between −20 and +60°C during charging which must be performed with a constant current at first, then at constant voltage until fully charged[9]. The figure also shows a second rechargeable accumulator, the EnerSi model 250 from Prollion, developed in order to supply the *Breitling Emergency II* watch, which includes a double radio transmitter.

Finally, there is an ultra-capacitor, Maxwell's PC 10 model[10], which we will return to later on. For the time being, we will note that its range of application extends from −40 to +70°C, and that in this range its ESR is constant and equal to 180 mΩ. Due to its low energy density and very high self-discharge, an ultra-capacitor cannot be considered as an energy source in the same way as a non-rechargeable battery. However, it will play a vital role in temporary storage, within the structure of a system for energy management based on harvesting ambient energy; this is why we have chosen to deal with them here.

Figure 2.1 allows us to verify several qualitative rules:

− non-rechargeable batteries are the components which present the highest levels of energy density, but are not able to provide the levels of power density of other technologies;

− rechargeable accumulators are more restrained than batteries in terms of temperature ranges, especially for the recharge phase;

---

7 This battery was used as the energy supply for the robot Philae which landed on the Churyumov–Gerasimenko comet in 2014 during the mission of the Rosetta probe.

8 The technical manual for this product does not give any information, but the self-discharge of a lithium accumulator can be higher than 1% per month.

9 The manufacturer recommends consulting them before use in the lower reaches of the temperature range, either for charging or discharging.

10 This is also a high-quality component that we selected from a range of other references during environmental tests. It was used as part of an equipment for the Mars exploration robot Curiosity. Unfortunately, it is no longer available commercially in parallelepipedic packaging.

– ultra-capacitors are the most efficient not only in terms of power density, but also possess low levels of energy density (which incidentally renders them less dangerous). Because of this, they are associated with electrochemical accumulators, in electrical vehicles, for example: they provide the power (acceleration), while accumulators provide autonomy (range).

However, it remains true that once a non-rechargeable battery is empty, regardless of how efficient it is, it will have to be replaced in order to maintain the proper functioning of the wireless sensor network. This is one of the main disadvantages of this system and for that reason, the creators of wireless sensor networks are considering harvesting ambient energy.

In the even more specific context of aeronautics, there are other criteria which argue in favor of avoiding, where possible, the use of electrochemical energy storage. Above all, this relates to the accessibility of nodes which are energetically autonomous: if several hours of disassembly and reassembly are required in order to access a node of a sensor network just to change a battery, the corresponding economic model is not particularly convincing. A second, somewhat paradoxical reason is the result of the high energy density of modern systems for electrochemical energy storage: approximately 1.7 MJ/kg for the LSH 20 model (see Figure 2.1) compared, for example, to the energy density of an explosive such as TNT, which is 4.6 MJ/kg[11]. The danger of electrochemical energy storage is clear, should the chemical reaction lose control due to extreme conditions of use or a manufacturing defect. The misfortunes encountered by Boeing in 2012 and 2013 with several non-contained partial combustions of new lithium accumulators installed in its recent *787 Dreamliner model* are a striking example of this[12]. Given that the cause of these accidents has still not been identified, the precautionary measures taken (a new layout of the battery's components, a new circuit charge, the use of a steel cover, smoke extraction ducts[13], and limiting the amplitude of charges and discharges) have led to a loss in the gain brought by the lithium technology in terms of energy density compared to the older, Ni-Cd technology (0.2 MJ/kg). In order to illustrate

---

11 There are storage methods which are even more energetic, but in order to release their energy, they require a chemical reaction with air, such as lithium-air accumulators (9MJ/kg). Jet fuel (JET A1) has an energy density of 43 MJ/kg.

12 Here, we would invite the reader to consult the article (unsigned) "Lithium-ion batteries: still at full power", in the review *Aircraft Technology*, number 135, April-May 2015.

13 In total, 84 kg of excess weight per plane (two batteries).

this point relating to the safety of accumulators exhibiting high energy density, we would invite the reader to examine the photograph in Figure 2.2 which shows the lithium battery developed by Saft for Airbus and the care which has gone into the packaging. The total weight of this battery is around 30 kg.

As a result, based on practical experience, for the remaining of this volume we will only consider the retrieval of ambient energy associated with the electrostatic storage of energy.

Figure 2.2. *Saft Lithium-ion battery for Airbus 350. From 2016 onward, each new Airbus 350-900 will come equipped with four of these. With the kind permission of Saft. © Stéphane Maréchal, Saft*

2.3. General points relating to the retrieval of ambient energy

2.3.1. *The structure of a system based on the harvesting of ambient energy*

Figure 2.3 shows the typical structure of an energy management system designed for supplying a node of a wireless, non-battery sensor network. The only aim of this illustration is to introduce a few key points and it can certainly be adapted for use depending on the design brief and how it is being used.

The input for such a system is composed of an exterior energy source, found in the environment around the node: solar or artificial light, mechanical vibrations, heat flows, electromagnetic waves, etc. The role of a transducer will be to transform this primary mechanical, optical or electromagnetic energy into electricity. Aside from the transduction operation, it is often essential to move onto a preprocessing of the energy from this stage onward:

– for example, in order to perform a rectification of the alternative output of the transducer (in the case of a transducer processing mechanical vibrations);

– in order to carry out impedance matching, thus maximizing the transfer of energy;

– in order to limit the electrical voltage applied to the storage ultra-capacitor.

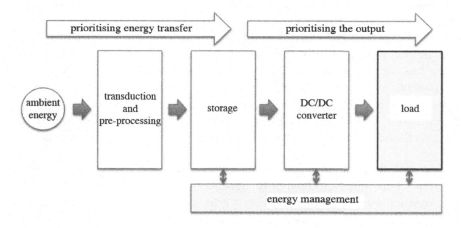

Figure 2.3. *Typical structure of an energy management system. Energy is taken from the environment around a node of a wireless, non-battery sensor network. For the remainder of this volume, storage will be considered as exclusively electrostatic*

At this level, the priority is to maximize the transfer of energy, which explains the function of the impedance matching explained above. Of course, the efficiency in this case is theoretically 50%[14] but at this stage, this modest value is meaningless, given that the priority is to obtain and store as much energy as possible.

---

14 The electrical energy obtained is equal to the electrical energy lost in the transducer.

The following stage is the storage stage, considered as being exclusively electrostatic – this is therefore a (super) capacitor – for the remaining of this volume. The following section will examine this in greater detail.

In Figure 2.3, a DC/DC convertor is downstream from the storage. Essentially, unlike electrochemical batteries and accumulators where there is little variation in the voltage so long as the energy reserves are not spent, the voltage at the terminals of capacitor follows the well-known law:

$$u = Q / C \qquad\qquad [2.1]$$

C is the capacity of the capacitor (in F), Q is the electrical charge stored (in Cb) and u (in V) is the difference in potential between the terminals of the capacitor. C will depend on the structure and size of the component and will remain constant for a given component. As a result, the potential difference will vary proportionally depending on the electrical charge stored. The aim of the DC/DC convertor is therefore to generate a constant voltage at its output, independent of the state of charge of the capacitor, and adapted to the biasing of the load. In practice, this load is composed of circuits and components necessary for carrying out the functions of nodes of the sensor network (measurement, signal processing, transmission, etc.) The output of the convertor, as well as the concept of the entire load, should maximize the overall efficiency, making best use of the energy obtained by the capacitor.

The last function mentioned in Figure 2.3 is the *energy management* function. This function is highly dependent on the application and the precise structure of the system which is only shown in a non-specific manner in Figure 2.3. This function, which can be partially implemented by the specific analog components associated with the capacitor, the storage and the convertor and by the digital circuits present in dedicated circuits, is able to carry out the following tasks:

– managing the independent awakening of the set, when the energy reserves are empty, and the electronic circuits are not polarized, once the transducer is ready to provide electrical energy again, given that conditions around the node are once more favourable;

– to take into account the state of the energy reserves in order to prepare the start of operations (such as the transmission of a radio frame) with the guarantee of being able to manage the operation until its completion.

2.3.2. *Justification and sizing of storage*

The requirement of a component for storing energy can be justified using two non-exclusive examples that we will now discuss[15]. Figure 2.4 shows the typical consumption of a node of a sensor network, from awakening to steady state in which the result of measurements carried out is regularly transmitted by radio.

The instantaneous power consumed varies considerably: there is an initial peak during the node's initial awakening[16], corresponding to an interim period during which the circuits and components are polarized, but are still not in their permanent state. Once this permanent state has been attained, consumption will drop and level out; however, each radio emission (high consumption, and assumed here to be activated intermittently in order to save energy) entails an increase in this level, itself preceded by an interim period of overconsumption which is also associated with the awakening of radio circuits.

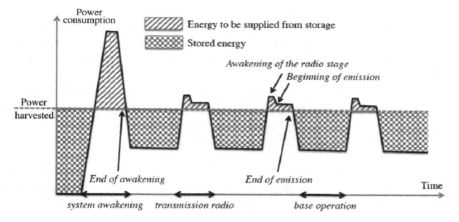

Figure 2.4. *Typical consumption of a node. The instant power consumed depends on the functioning phases, while the power harvested is assumed to be constant*

---

15 A third implicit justification is to not directly call upon the energy transducer by current draw. There is often a relatively high level of impedance with the transducer. Storage makes it possible to maintain the condition for impedance matching at output of the transducer and to thus prioritize the transfer of energy (see Figure 2.3).

16 After the biasing application, integrated circuit transistors will not immediately be in their logical final state (ON or OFF) and current draws can take place, for instance with simultaneous conduction.

In this figure, the electrical power harvested is assumed to be constant, which does not limit the argument: it is, however, assumed to be less than the peak power required by the load. This situation is a good indication of what prevails in terms of the retrieval of ambient energy: restrict the volume (and therefore the weight) of the energy transducer to the necessary minimum. In practice, the average power obtained is often lower than the peak values of power consumed. However, it is possible to deal with this situation provided that there is a storage system which enables the energy stored to "even out" the peak values of power consumed by the load. If we refer to Figure 2.4, it is necessary for the patterned area to be larger than the cross hatched area (the areas are energy images).

While the storage is guaranteed by a (super) capacitor, calculating its capacity will follow the calculation below. It is important to bear in mind that the energy E (in J) stored in a capacitor with a capacity of C showing a potential difference of u between its electrodes is given by:

$$E=1/2Cu^2 \hspace{3cm} [2.2]$$

If we fix u_{max} the maximal potential difference (in V) supportable by the capacitor (around 5 V for a supercapacitor[17]), u_{min} the minimal voltage (in V) as an input of the DC/DC convertor guaranteeing this functions properly (obtaining nominal regulated voltage as an input of the load), $\eta(\%)$ the efficiency of the convertor and E the energy (in J) to be stored, the capacity C (in F) of the capacitor must verify the equation:

$$E = \eta \; 1/2C \; [u_{max}^2 - u_{min}^2] \hspace{3cm} [2.3]$$

Based on our experience, the necessary capacity values are in practice often between several hundred millifarads and a few dozen farads, values which require the use of supercapacitors.

A second configuration involves the use of a storage component: for this, unlike with Figure 2.4, the ambient energy source is intermittent, giving considerable variations in the power harvested. This situation is represented schematically in Figure 2.5. This situation is quite similar to obtaining solar energy, with a period of repetition which follows a 24-h cycle. It should also be noted that, in this instance, in phases where the value is greater than zero,

---

17 In reality, this value is normally somewhere between 2.5 and 2.7 V, but manufacturers often link two supercapacitors in series in order to double the maximum working voltage.

the power obtained must be far higher than the average power consumed by the load. Indeed, the clipping of the peak power transients in Figure 2.4 is therefore implicit.

Sizing of the storage can once more be managed using formula [2.3], except in instances – such as ours – involving the use of electrostatic energy storage, where it is essential to take into account the self-discharge once the temporal scale of intermittence exceeds several hours. Unfortunately, as will be explained in the following section, the nonlinear nature of this phenomenon does not lend itself to writing a simple mathematical formula; in practice, the designer will employ an oversizing of the storage by a large margin.

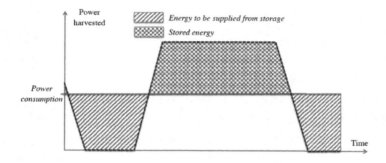

Figure 2.5. *Power obtained from an intermittent ambient source. In the interest of simplification, the power consumed is assumed to be constant*

2.3.3. *A few points relating to supercapacitors (also ultra-capacitors)*

Given that it has already been mentioned several times, for the remainder of this volume, we will only consider the solution of electrostatic energy storage; as is explained in the previous section, the quantity of energy to be stored requires capacity values that only supercapacitors are able to provide. In the following section, we will provide a concise explanation of the characteristics and performances of these relatively new systems[18].

---

18 They have, however, been present (and therefore certified) in certain airliners, such as the Airbus A380 in which they are used for opening certain emergency exits: operating voltage 2.5 V, and capacity in the region of 100 F per system.

The capacity of a capacitor is traditionally given by formula [2.4] below:

$$C = \varepsilon S/e \tag{2.4}$$

C is the capacity (in F), S is the equivalent surface of the electrodes in view (in m^2), e is the distance between electrodes (in m) and ε is the permittivity of the dielectric or the electrolyte[19] positioned between the electrodes (in F/m). Dielectric capacitors only make it possible to access low capacity values (up to microfarads) and are not polarized. For higher capacity values (up to several hundred millifarads, even higher), it is necessary to use electrolyte capacitors, which, however, have the double disadvantage of being polarized, heavy and cumbersome (with working voltages which are nevertheless higher than the supercapacitors presented below).

In order to access the farad range with limited weights and below reasonable volumes[20], it is necessary to select supercapacitors. These will possess the following two specific characteristics:

– porous nanostructured electrodes which significantly increase the equivalent surface S of the electrodes;

– an electrolyte which enables ionic conduction, and develops a double layer of electrical charges opposed to each of the electrodes[21], thus reducing the equivalent distance e appearing in equation [2.4].

Remembering that the mobilities of anions and cations in electrolytic solutions are often different, the electrodes are adapted, one to cations and one to anions, and as a result, if the user wishes to exploit the nominal characteristics of the component, they must respect a polarity. Finally, this will mean that the distance between charges is very weak, which limits the maximal potential difference at the terminals of the electrodes to between 2.5 and 2.7 V.

---

19 A dielectric is an insulator, in other words, a material in which electrical charges cannot move large distances. An electrolyte is a conductive substance in which the electrical current is carried by ions (anions and cations).

20 Without wanting to make this a general rule, the use of a supercapacitor appears to give a gain in energy density between 10 and 20, given that the gain is higher for mass density than with volumetric energy density.

21 For this reason, supercapacitors are often referred to as Electrochemical Double Layer Capacitors (EDLC).

Because no chemical reaction is involved, the storage of energy by supercapacitors can be considered as relevant within the realm of electrostatic energy. This characteristic enables supercapacitors to age less during periods of charge and discharge: for example, a drift of 20% in the capacity value for 500,000 cycles, for the Maxwell PC10 in Figure 2.1, while the Saft MP 144350 rechargeable battery appearing in the same figure will have lost 20% of its capacity (in Wh) after only 500 cycles (manufacturer's data). This efficiency argues in favor of the use of supercapacitors with wireless sensor networks that have been abandoned (due to difficulty of access) in the complex mechanical structure of an aircraft. Figure 2.6 gives an example of this.

Figure 2.6. *Example of a circuit board conforming to the structure shown in Figure 2.3 and using two Maxwell BCAP0003 3.3 F supercapacitors in series (authors' photograph)*

However, one area where supercapacitors are less effective is in terms of self-discharge. The phenomenon is nonlinear and cannot be modeled using a constant leak in resistance in parallel to the supercapacitor: while initially the discharge is exponential, the time constant increases over time [KOW 09]. Furthermore, the dynamic of self-discharge will depend on the temperature, the initial charging voltage, the nature of the charge (intensity of the current) and the length during which the initial voltage was maintained at the terminals of the fully charged supercapacitor. Furthermore, the authors noticed a certain optimism in user manuals when this point is raised. In order to set the ideas, for four different references of supercapacitors with values of 1.2 and 1.5 F, charged at 5.5 V under a current of 5mA, at the end of 3 h the authors were able to measure a residual energy, following the references, from 55–90% of the initial stored energy. It is therefore essential, especially during the sizing of an application for which energy is intermittent (see Figure 2.5) to precisely characterize the storage phase in the relevant conditions for use (temperature and charge characteristics).

Finally, it should also be noted that there are limits to a supercapacitor in terms of energy density when the desired autonomy is large and the charge consumption is high (as shown in Figure 2.5). Figure 2.7 illustrates this in an extreme case for which, aside from a high-capacity (90 F), specification imposed a reduced thickness, leading to a proliferation of small, flat supercapacitors, covering a large surface.

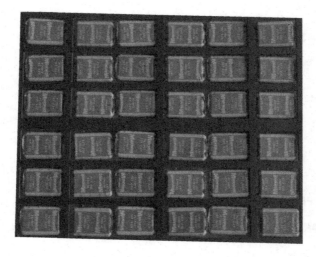

Figure 2.7. *Matrix of 10 F Maxwell PC10HT supercapacitors in 2S18P association giving an equivalent capacity of 90 F under 4.4 V maximum. Maintaining a reduced thickness leads to a large surface devoted to storage (authors' photograph)*

2.4. Ambient energies and associated transducers

The concept of harvesting ambient energy present in the environment surrounding the system to be supplied requires the interception of external *flow* of energy: a system immersed in an isothermal environment, or subject to constant acceleration, will not be able to extract anything from its environment. Given this, and restricting ourselves to aeronautics, what types of energy flow can we contemplate? Here is a list, which is probably largely complete:

– light energy, from outside the aircraft (solar flux) or inside (artificial lighting);

– temperature gradients, between the exterior and the interior of the aircraft, or developing through exothermal equipment;

– mechanical energy: distortions and vibrations in the structure, acoustic noise, variations in pressure due to altitude, relative wind effects;

– electromagnetic energy: radio waves, induction produced by the passage of strong currents in the power network.

In Chapter 4 of this volume, we will illustrate several scenarios for using some of these energy sources. In the current chapter, we will present the sources which seem to be the most relevant. Aside from hypothetical considerations, one of our selection criteria was the commercial availability of reasonably reliable transducers. We will put ourselves in the position of a user and not a designer of these transducers.

2.4.1. Harvesting light energy

The photovoltaic effect discovered by Antoine Becquerel in 1839 is linked to the generation of free electrons in a semiconductor subjected to a photon flux. For a long time considered to be a laboratory phenomenon, due to the progress made in microelectronic technology the first industrial applications of photoelectricity were seen in the 1960s in space exploration. This soon spread to various other applications.

Specifically, the transfer of energy between a photon and an electron in the valence band of a semiconducting material will cause it to move in the conduction band, thus generating an electron-hole pair. If this pair is generated in the space charge region of a PN junction (for a photodiode), the electrical field will sweep the hole and electron in opposite directions, thus generating an external electric current if the photodiode is linked to a load. Figure 2.8 presents a summary of this by showing the typical current-voltage curve of a photodiode. If the operating point of the diode is situated, by external biasing, on the quadrant identified as a "high impedance receiver", the diode will then be in photoresistance mode. If on the quadrant identified as a "generator", the photodiode will supply the electrical energy to an eventual load.

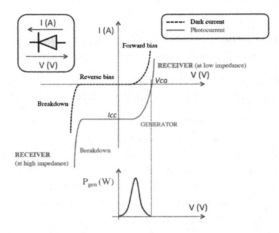

Figure 2.8. *Current/voltage response of a photodiode. The cartridge on the left indicates the conventional direction of the voltage and the current (passive sign convention). In generator mode, the power P supplied to the load is also determined as a function of the bias*

In Figure 2.8, the power versus voltage supplied by the photodiode is also outlined: between a situation where the diode is in short-circuit, supplying a current I_{CC}, and the case of an open circuit for which the diode presents a potential difference V_{CO}, at its terminals two situations for which no power is supplied, there is a point of biasing for which the power transmitted P_{gen} to the connected load is maximal: this is known as the maximum power point or MPP. For obvious reasons, this situation will be sought after in order to maximize the transfer of energy (for impedance matching). Yet, this operating point will vary depending on two parameters: the operating temperature and the lighting[22]. Here, it will be useful to implement a procedure known as maximum power point tracking or MPPT. There are a number of algorithms: dynamic tracking by successive trials (*hill climbing*) or the regular measurement of V_{CO} (or I_{CC}) and the application of a

22 Particularly for weak levels of radiation, a situation which cannot be ignored in the context of harvesting ambient energy.

coefficient of proportionality in order to define the output current and voltage applied to the load. However, this second method requires a preliminary characterization of the photodiode. Finally, in order to simplify electronic circuits, it is possible to avoid any dynamic pursuit of the maximum power point by determining *a priori* a load line providing a good compromise.

Solar radiation varies from 130 mW/cm^2 in space to around 100 mW/cm^2 on the ground in good weather, with the sun at its highest point. In cloudy weather, it will drop to around 10 mW/cm^2. With artificial light (here there can of course be significant variations), luminous flux will be lower than 1 mW/cm^2. With light at 100 mW/cm^2 and a silicon photodiode, at the maximum power point, the delivered voltage will be around 475 mV, which is relatively low, and this explains why photodiodes are often associated in series and in parallel in order to increase the voltage and the current, forming a photovoltaic *panel*.

However, the spectrum of sunlight, even on the ground, is large and extends from infrared (high wavelengths and low energy photons) to ultraviolet (short wavelengths and high energy photons): schematically, a low energy photon (lower than the band gap) is not absorbed, a high energy photon allows an electron to pass into the conduction band; if its energy is much higher than the width of the band gap, the excess energy will be transformed into heat. However, the efficiency of a photodiode decreases very quickly with the temperature. So-called multi-junction cells, associating several PN junctions of different semiconductors, are therefore used in order to control these two effects: to begin with, photons will cross a large gap semiconductor, a semiconductor that will absorb most of the energy, ensuring that the cell does not overheat, then the lowest energy photons (lower than the first gap encountered) will be absorbed by semiconductors presenting gaps which get progressively smaller. In other words, UV rays and the top of the visible spectrum will be absorbed first, while the bottom of the visible spectrum and infrared will be absorbed later in the multi-junction cell.

When it comes to making a choice between the different options, it will be useful to know that:

– the peak power of a cell, as listed by the manufacturer, will be for a temperature of 25°C, with an illumination of 100 mW/cm^2 and discharging on an adapted load;

– the efficiency of a silicon, mono-junction cell in 2015 is around 27% for the most efficient[23], a similar value to the theoretical maximum efficiency (31%);

– the efficiency of a multi-junction cell in 2015 is around 37% for the most efficient and reaches 44% when equipped with an optic concentrating the radiation;

– the efficiency will fall by roughly 0.5% per degree above 25°C;

– finally, the versions on flexible substrates (which can be adapted to non-flat surfaces) possess efficiency values far lower than their equivalents on rigid substrates (see Figure 2.9).

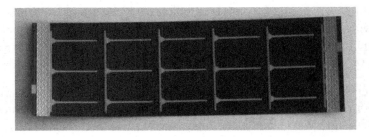

Figure 2.9. *PowerFilm© MP3-37 flexible cell. The peak power given is 150 mW for a surface of 41.6 cm², a calculated efficiency roughly 3.6%, lower than rigid substrates (authors' photograph)*

In concrete terms, for applications where sunlight is available, in good conditions with direct sunlight, *electrical* power values of several dozen milliwatts per cm² are possible, that is with a level of power which makes it possible to supply so-called commercial low consumption modules, combining a microcontroller, a sensor and an RF front end[24], i.e. the full array of functions required by a wireless sensor networks. In cloudy conditions, when the sun is low on the horizon (early and late in the day, winter, higher latitudes), this power decreases very quickly. It will be the same with artificial light for which conditions can be unsuitable for sufficient levels of energy retrieval (for a reasonably sized photovoltaic cell).

23 The efficiency values given here come from data published by the *National Renewable Energy Laboratory*, readily available on the Internet.
24 For example, the Jennic models from NXP which have an average level of consumption (except when starting-up) of a few hundred microwatts.

Remaining within the realm of aeronautics, light energy must be considered as an intermittent ambient energy, whether this is sunlight outside of the aircraft or artificial light inside the cabin (Figure 2.10), which only supplies a relatively small amount of light energy. The possibilities are therefore limited, although there are niche applications, such as those examined in more detail in the final chapter of this volume.

Although permanent innovations are constantly being made in the realm of photovoltaic cell technology, the industry is today in a position to offer well-developed, high-quality products.

Figure 2.10. *Measuring light in the cabin of an Airbus A330 (carried out by the authors): the luminous flux is intermittent and relatively low (©Airbus)*

2.4.2. Using thermal gradients

2.4.2.1. General points relating to thermoelectricity

The connection between thermal energy and electricity is provided by two physical phenomena:

– pyroelectricity, a phenomenon in which a material[25] subject to a temporal variation in temperature generates an electric charge;

---

25 Pyroelectric materials are always piezoelectric, but the opposite is not always true.

–thermoelectricity, a phenomenon in which a material subject to a temperature gradient, generates an electric charge.

Pyroelectricity will not be explored in the following volume, because in order to generate significant power values, there must be frequent variations in temperature (the power generated is proportional to the frequency of the phenomenon). However, this is a somewhat rare occurrence, and something which we have not personally encountered. We should though refer to laboratory transducers which, subjected to a permanent gradient, will create a temporal thermal oscillation through the use of a deformable pyroelectric structure, in contact alternately with hot and cold surfaces linked to the gradient, similar to a bimetal [RAY 11].

Thermoelectricity[26] makes use of the well-known *Seebeck*[27] effect, the appearance of a potential difference ΔV *(V)* between the extremities of metallic or semiconductor material, with the extremities subject to temperature difference of ΔT (°K). In the absence of a gradient, free carriers will spread out in a uniform manner, but when there is a difference in temperature, the carriers on the warm side will display more kinetic energy than on the cold side, and circulate toward this side, creating an electrical field. Equation [2.5] makes it possible to quantify the phenomenon using the Seebeck coefficient S (V/°K)[28] of the material considered (*S* here is considered as being independent of the temperature, a hypothesis which is justified in our context because ΔT is sufficiently weak).

$$\Delta V = S \,\Delta T \qquad\qquad\qquad [2.5]$$

The Seebeck effect is widely used for measuring temperature with thermocouples (incorporating two metals or alloys, with different thermoelectric characteristics, welded together at one of their extremities, with the coefficient S in equation [2.5] therefore becoming a *differential* Seebeck coefficient).

It should be noted that when a conductor material is subjected to an electric current, exchanges in thermal energy take place which lead to the

---

26 There will be thermoelectric effects with all conduction materials, apart from superconductors [SPIE 15].
27 Named after the scientist who discovered it, Thomas Seebeck, who brought it to light in 1821.
28 The unit in S is in practice µV/°K. Furthermore, S is often denoted as α.

opposite phenomenon from the Seebeck effect: the *Peltier*[29] effect, used for cooling in some applications. We will discuss the Peltier effect at a later stage, as an unavoidable "parasite" effect of retrieving energy through thermoelectricity. Finally, there is also the *Thomson* effect, which is characterized by the lack of a need to couple different materials together in order to demonstrate electricity, as is the case with the two previous effects. In terms of energy retrieval, however, it can be ignored [SPIE 15].

With regard to generating energy by thermoelectricity (for the remainder of this volume, we will only consider those power ranges necessary for supplying wireless sensors[30]), pairs of different materials are associated thermally in parallel, i.e. they are all subjected to the same thermal gradient. Electrically, they are associated in series and in parallel in order to obtain voltages and currents from the thermogenerator which are compatible with the energy management circuits downstream.

The performance of a thermoelectric material is traditionally evaluated using a figure of merit Z defined as follows [IOF 57]:

$$Z = \sigma \, S^2 / \lambda \qquad\qquad [2.6]$$

where S is the Seebeck coefficient of the material, already defined in equation [2.5], σ is its electrical conductivity (Ω/m) and λ is its thermal conductivity (W/m. °K). For initial analysis (more detailed considerations will be provided in the following section), a thermoelectric material will therefore be much more efficient with a higher level of electric conductivity and higher Seebeck coefficient (and therefore a weak level of internal electric resistance enabling the transfer of high electrical power on a load) and a weak level of thermal conductivity (making it possible to maintain a significant thermal gradient at its terminals when inserted into a given environment).

---

29 Discovered by Jean Peltier in 1834.

30 At power levels of several hundred watts, far higher than the levels considered in this volume, thermoelectric generators have been used since 1961 in space probes and planetary robotics in order to generate energy using the heat produced by dissociating radio isotopes: 9 years after it was launched, the *New Horizons* probe which explored Pluto in 2015 had at its disposal 90 W of power which was "thermogenerated" using 11 kg of Plutonium-238 dioxide. Thermogenerators are sometimes used in ground transport in order to retrieve thermal energy from exhaust systems. Studies are also being carried out in aeronautics for high-power applications, but the weight could potentially be a handicap.

Figure 2.11 shows a qualitative example of the evolution of the above parameters as a function of the concentration of free carriers. It summarizes the following physical phenomena:

– the concentration of free carriers is strongly linked to the three parameters in equation [2.6];

– unfortunately, the Seebeck coefficient develops in the opposite direction from the electrical conductivity;

– electrical and thermal conductivity develop in the same direction for higher concentrations of carriers, due to the contribution of these carriers in the transfer of heat;

– the figure of merit displays a maximum for average values of carrier concentration, situated between those of insulators and metals, similar to the levels of concentration encountered with a semiconductor which has been significantly doped.

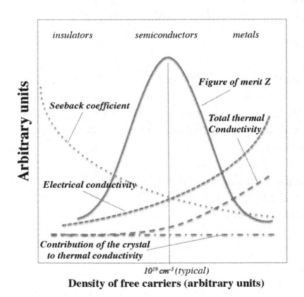

$10^{19}\,cm^{-3}$ (typical)
Density of free carriers (arbitrary units)

Figure 2.11. *Qualitative changes in electrical and thermal conductivity of the Seebeck coefficient and the figure of merit Z as a function of the density of the carriers*

The best results are therefore obtained using semiconductor materials[31], given that each couple is formed of a doped bloc P and another doped N (see Figure 2.12). Given that the voltage delivered by a single thermocouple is weak (around a few hundred $\mu V/°K$), several hundred thermocouples tend to be associated (see Figures 2.12 and 2.13) in order to form a *thermopile* or *thermogenerator*.

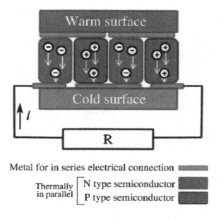

Figure 2.12. *Typical structure of a traditional thermogenerator with a vertical structure. In this illustration, the semiconductor blocs P and N are electrically in series in order to increase the output voltage. Associating them in parallel is also possible in order to increase the current. This illustration demonstrates why it is necessary to use two types of doping*

Figure 2.13. *Thermogenerator (3 cm × 3 cm) showing the metallic interconnections (clear sections) between semiconductor blocs (non-visible). Bringing hot and cold bodies into a thermal contact thermogenerator should be electrically insulating (Hi-Z Technology Inc. model, authors' photograph)*

---

31 It was due to the development of the physico-chemistry of semiconductors in the 1950s that thermoelectricity was really able to develop, given that the conversion efficiency with metals had previously been limited to roughly 1% [BIR 13].

Pairs formed in this way, as well as metal platings which enable the electrical association of elements, are taken into a sandwich between two ceramic plates, a material which is both electrically insulating and a relatively good conductor of heat[32] (Figure 2.14). Two wires make it possible to connect the group of circuits downstream for energy processing. The efficiency is modest (somewhere in the region of between 5 and 10%) but will increase with the thermal gradient. Subjected to a gradient of 10°K, this type of thermogenerator typically delivers 100 mW for a surface of 25 cm^2 [MON 14a].

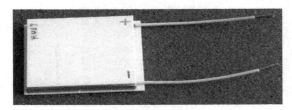

Figure 2.14. *Traditional appearance of a thermogenerator (5cm × 5cm) with semiconductor blocks (non-visible) inserted between aluminum oxide plates (preproduction from the Kelk Komatsu company, authors' photograph)*

Furthermore, when using semiconductor materials, the performance of a thermogenerator is highly dependent on the concentration of free carriers, and therefore the nature of the semiconductor used as well as the temperature. The majority of thermogenerators available commercially[33] operate up to roughly 200°C, some up to 250°C and in exceptional circumstances as high as 280°C. The majority use bismuth telluride (BiTe$_3$)-based semiconductors. It should be noted that there are new thermogenerators designed for use at very high temperatures, using silicon alloys (hot side up to 580°C) or requiring metallic vacuum packaging in order to avoid any oxidization, with silicon germanium alloys (range of 800°C)[34].

---

32 An important characteristic in order to avoid too large a part of initial thermal gradient that we aim to use is not found at the edges of these plates, without having an effect on the generation of energy.
33 The commercial range is large, some companies (for example, in Japan) were founded in the 1950s, and this can be explained in part by the historical (and long term) use of these systems for use in cooling (Peltier effect).
34 See, for example, the products designed by the French start-up *HotBlock OnBoard*.

The search for more efficient performance generally involves studies on the materials designed for partially decorrelate thermal and electrical conductivity (see Figure 2.11) by reducing the component of the thermal conductivity carried by the phonons (responsible for heat conducted by the crystal semiconductor and not by the free carriers). In other words, it aims to reduce the floor level of the thermal conductivity (the initial level of the curve in Figure 2.11) essentially due to the crystal in the absence of a significant number of free carriers. This method involves working on the nanometric structure of the semiconductor [BIR 13, SPIE 15], knowing that it is not necessary for the semiconductor to be monocrystalline (it can be polycrystalline) as is the case with microelectronic technology. One method is to include atoms with a weak connection to the crystal in crystalline grains, thus limiting the spread of phonons. The commercial applications of these studies remain to be seen.

Although they are probably less widespread, other technologies do exist:

– some are based on collective processing (microelectronic) of N and P regions, followed by a wafer bonding (for example, products from Micropelt, GmbH, Nextreme Thermal Solutions Inc. and Thermo Life ®) making it possible to foresee miniaturized systems in the future. Z figures of merit are relatively weak, but given that far more N and P elements can be integrated per surface unit, output voltages can be high even at low gradients, which enables an early wake-up of the associated electronics;

– flexible substrates (for example, products from greenTEG);

– in laboratories, MEMS technology is used in order to manipulate the thermal gradients [SIO 12] with the aim of making them horizontal within the thermogenerator and thus to control the thermal conduction surfaces and to decorrelate electrical and thermal conductance.

Given that the principle of converting thermal energy has been established, we will now turn to consider the integration of the thermogenerator in its environment.

2.4.2.2. *The integration of a thermogenerator in its environment*

Unlike a photovoltaic cell which is delivered *ready to use*, a thermogenerator must be integrated into a mechanical system adapted to its environment[35], a mechanical system which will therefore be involved in the transfer of thermal energy. The way in which this mechanical system is set up will change on a case-by-case basis. Furthermore, the theoretical aspects mentioned above refer to a thermogenerator perfectly paired with heat reservoirs which are supposed to be thermostatic, on which the thermogenerator therefore has no impact, which is not necessarily the case.

Within the framework of this volume, we will not go into too much detail on the efficiency of a thermogenerator and how it interacts with its environment, because this type of study is – as far as the authors are concerned – of little practical use. However, we will explain several aspects which, at least from a qualitative point of view, are of significance. For further information and a detailed, pedagogic approach to these complex phenomena, we would direct the reader to the doctoral thesis of Romain Monthéard [MON 14a][36].

It would be erroneous to suppose that the maximization of electrical power[37] coming from a thermogenerator is carried out via simple electrical impedance charge matching and by maximizing the gradient. On this last point, if it were true, a perfect insulating thermogenerator would be preferable; however, in this case, there would be no thermal flux to harvest and no energy to retrieve. The thermogenerator supplying electrical energy must therefore interact thermodynamically with its environment:

– even when the thermogenerator is not supplying a current, it will have an impact on the existing thermal balances previously in place because it has neither zero nor infinite thermal conductivity;

– when a current is supplied to the charge, the Peltier and Joule effects interfere with the state of balance attained in the absence of a current (therefore, the thermogenerator cannot be represented by a Thévenin model since the current released will have an influence on the electromotive force, the average temperature of the thermogenerator and therefore its thermal resistance);

---

35 Some manufacturers offer preassembled assessment kits.
36 Full text available online on the open archives server HAL.
37 It should be noted that this optimization, and not that of the efficiency, is to be considered.

– part of the thermal gradient is developed in the heat reservoirs and the various thermal contacts present in the mechanical set-up of the thermogenerator.

In summary, the maximization of the electrical power therefore requires a double impedance matching, which is not only electric but also thermal in order to maximize the flow of heat in the thermogenerator, given that electric and thermal mechanisms are linked. For further information, we direct the reader toward the various equations given in [MON 14a] in order to fully make the most of their system. However, based on the authors' experience, this kind of optimization is difficult for the following reasons:

– there is often little information about the thermal environment outside of the thermogenerator, and it is difficult to model (see below);

– the thermal performance of the mechanical system in which the thermogenerator is located (see below) is also uncertain: for example, the thermal resistance of the different contacts is dependent on the state of the surfaces, on the nature of the thermal grease used in order to fill microscopic gaps and finally on the mechanical pressure used during assembly.

– the technical manuals for thermogenerators are often brief and imprecise in terms of the thermal and electrical properties of their components[38]; it should be noted that efforts are being made to formalize the procedures for characterization and to identify reference materials [PER 13];

– the design brief sometimes discriminates a lot between parameters which are not directly linked to the efficiency of the thermogenerator (for example, dimensions, packaging, maximum temperature for use and range of output voltages) and requires making a choice from a limited number of references, depending on the power value given by the manufacturer, for a gradient which has often been only roughly estimated by the user.

2.4.2.3. Thermogeneration and aeronautics

Inside an aircraft, there are various thermal gradients; all internal sources of heat generate them, including (but not limited to) electrical and electronic bays, actuators, fluid lines, motors and brakes.

---

38 Simply measuring the conductivity σ cannot be effectively carried out since standards such as IEC 468 and DIN 50431 were developed for metals.

Thermal gradients also gradually develop between the exterior and the interior of the aircraft (depending on the altitude) and therefore between internal surfaces and the air in pressurized and heated compartments. Some gradients are more or less permanent, some are temporary, some have values which are largely predetermined and others are random. It should be noted at this stage that we will not deal with those caused by the proximity between electric and electronic equipment, although permanent and foreseeable, because the proximity of the electrical network of the aircraft weakens the relevance of the concept of energy harvesting[39].

The authors of this volume were behind one of the first developments in aeronautics for retrieving energy through thermoelectricity using phase changing materials [BAI 08][40]. The aim is to use the decrease in outside air temperature (typically between −50 and −70°C[41]) when an airliner reaches its cruising altitude, the air outside constituting the *cold surface*. The *warm surface* can be made of any part of the aircraft's structure provided that an isothermal state has not been reached. The principle put forward in [BAI 08] is to prolong the existence of this temporary gradient by using a small (a few cm^3) reservoir of water on the warm side, which is used for its calorific capacity as well as the latent heat[42] of fusion/crystallization (see Figure 2.15). The gradient is only temporary, but it is also present during descent (then it will be in the opposite direction) and can affect any position on the aircraft, even and above all those areas which are furthest from any internal heat source. The use of a phase changing material (here water, presumed to freeze and thaw during a flight) was the result of an assessment carried out by the authors that found that thermal gradients between the

---

39 Not forgetting the constraint which is often put forward saying that "the energy harvester must weigh less than the power line it is replacing".

40 The concept of using phase changing materials was simultaneously and independently proposed by another team [SAM 11], grouping together EADS IW (today Airbus Innovation Group, AIG) a team which had been developing advanced prototypes as part of the StrainWISE program *Clean Sky*.

41 Air temperature decreases with altitude. The ICAO (the International Civil Aviation Organization) applies a standard of +15 °C at sea level and −50 °C at 10,000 m. The reality can be very different.

42 The latent heat of fusion recrystallization of water is very high, 330 J/g, used twice in our scenario so 660 J/g (ignoring the efficiency of the thermogenerator) compared with the electrical energy density of 1,700 J/g of the Saft LSH 20 battery (Figure 2.1).

internal surface of the aerodynamic sections of an airliner and the interior of non-pressurized and non-heated sections were too weak to supply wireless sensors by thermoelectricity (Figure 2.16). Only in the case of the passenger cabin are there sufficient gradients (much higher than 10°C).

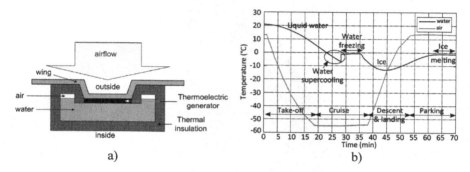

a)   b)

Figure 2.15. *a) Structure leading to the creation of a temporary thermal gradient between the outside air and a water reservoir, water being used as a phase changing material. The part identified as the "wing" corresponds to any structural element of the aircraft exposed to the outside air; b) Experimental change in air and water temperatures for a flight recreated in a climate chamber. The thermal gradient at the edges of the thermogenerator is the gap between the two curves; it will be inverted during descent [VAN 15]*

It was important when carrying this out for it to be independent of any heat source inside the aircraft. This allowed us to offer innovative structures in terms of the electronic circuits for the harvesting and management of ambient energy [VAN 15], but they suffered from a number of disadvantages. Chief among these is a high dependence on meteorological conditions, and, for example, an almost total loss of efficiency in instances of take-off in negative temperatures. Furthermore, the retrieval of energy only takes place during ascent and descent, and not while cruising, which is often the longest phase.

However, this first experiment illustrates a crucial point: the harvesting of thermal energy on an aircraft often happens (in every case during experiments carried out by the authors) by intercepting thermal flows between a solid body and air. We are not aware of any example of the implementation of thermoelectricity on an aircraft (for supplying wireless sensors) by inserting a thermogenerator between two solid bodies, or by using a heat pipe.

Figure 2.16. *Sensors installed by the authors for measurement on an Airbus A340-600. The thermal gradients which develop far from any heat source are weak, especially those shown here on a hatch made of a composite material (© Airbus)*

For the hypothesis of an installation close to a heat source, the thermogenerator will be, on one side, indirectly in contact (through the intermediary of a plate) with a solid wall and will exchange through convection with the air on its second side through the intermediary of a heat sink. An example of this is shown in Figure 2.17. A cross-sectional view of the main corresponding structure, as well as the different gradients, is given in Figure 2.18. In practice, the solid wall will be the heat source and the air will be the cold source. For the remaining of this study, only those exchanges by conduction and convection will be considered. In cases where walls are at very high temperatures, it is necessary to carry out an evaluation on the impact of the radiation on thermal exchanges.

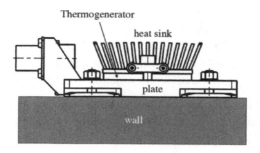

Figure 2.17. *An example of the assembly of a thermogenerator on a wall, including heat sink by convection*

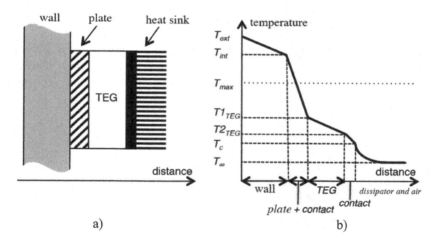

Figure 2.18. *Thermogenerator (TEG) carrying out simultaneous exchanges by conduction and convection: a) general structure and b) the appearance of different temperature gradients*

In this figure, the temperatures identified have the following implications:

– T_{ext} is the temperature of the wall far from the mechanical structure in which the TEG is inserted, and T_{int} is the temperature of the wall in contact with this same structure; if the wall is metallic (good conductor of heat) and thick, T_{ext} and T_{int} can be thought to be identical (isothermal case);

– T1$_{TEG}$ and T2$_{TEG}$ are the two temperatures of the two sides of the thermogenerator. These are what cause the thermal gradient ΔT of the thermoelectric effect. T1$_{TEG}$ is lower than T$_{int}$ due to the presence of a plate for the mechanical assembly of the thermogenerator and different non-ideal thermal contacts between the different assembly components;

– T$_{max}$ is the maximum use temperature of the thermogenerator as outlined by its manufacturer. In instances where the temperature of the wall is higher than T$_{max}$, a material which is an average conductor of heat can be chosen for the plate in order to intentionally decrease the value of T1$_{TEG}$;

– the difference in temperature between T2$_{TEG}$ and T$_C$ (which represents the surface temperature of the heat sink) is due to the non-ideal contact between the thermogenerator and the heat sink and to its thermal conductance;

– T_∞ is the air temperature far from the system.

When a steady state is attained, the same heat flux Φ (W) flows through the different elements of the system in Figure 2.17, and assuming unidimensional exchanges, it is then possible to write the following equations:

$$\Phi = K_p \left(T_{int} - T1_{TEG} \right) \tag{2.7}$$

$$\Phi = K_{TEG} \left(T1_{TEG} - T2_{TEG} \right) = K_{TEG} \, \Delta T \tag{2.8}$$

$$\Phi = K_c \left(T2_{TEG} - T_c \right) \tag{2.9}$$

$$\Phi = h.A \left(T_c - T_\infty \right) \tag{2.10}$$

With K$_p$, K$_{TEG}$, K$_c$, being the thermal conductivities (W/°K) of the plate and thermal contacts, of the thermogenerator and finally of the heat sink and thermal contacts; h represents the heat transfer (convection) coefficient of the heat sink (W°K^{-1}m^{-2}) and A is the area of the heat sink (m^2). If we are aware of the limits T_{int} and T_∞, it is very easy to calculate the intermediary temperatures, and therefore the gradient ΔT, and as a result the thermogenerator's output voltage using equation [2.5] or more realistically by using a characteristic given in the user manual for the thermogenerator.

However, it should be noted that this apparent simplicity can be misleading:

– thermal conductances are difficult to evaluate and the thermal conductance of the thermogenerator is often not given by the manufacturer;

– the model given above is only valid in a static system;

– generally speaking, convection is a phenomenon which is very empirically modeled retrospectively;

– boundary conditions are often uncertain, including the nature of the convection (natural or partly forced);

– in this model, the thermogenerator does not deliver any energy and from this point of view does not therefore interfere with the flux Φ.

However, this model does allow for an initial evaluation of the system's efficiency in terms of harvesting thermal energy, and therefore a rough sizing. Above all, it highlights the importance of maximizing the thermal gradient at the edges of the thermogenerator by minimizing the thermal contact resistance, avoiding *thermal bridges*[43] and finally by the fact that an efficient heat sink is required.

The vast majority of thermal heat sinks work by extending the surface and using fins positioned on a base plate. Rectangular fins are most effective, provided that they are positioned in the direction of the natural or forced convection. When this direction is unknown or variable (turbulent flow) cylindrical fins are preferred. The thermal conductivity of the material used for the heat sink must also be as low as possible: from this point of view, the use of pure copper is preferred.

One final point is specific to aeronautical applications: this is the rapid decrease in the thermal convection coefficient h with the altitude. At 3,500 m, the coefficient h is reduced by 25%[44], and by 40% at 6,000 m[45]. However, this worrying trend is balanced out by two other phenomena:

---

43 Caused, for example, by the system bonding the heat sink to the plate, a mechanism which puts pressure on the thermogenerator in order to reduce the values of thermal conduct resistance (see Figure 2.17).
44 According to the document *How to select a heat sink* by Seri Lee, available online.
45 According to the document *Adjusting Temperatures for High Altitude* from the online review *Electronics Cooling*, 1st September 1999.

– in practice, the plate in Figure 2.18 often constitutes the hot surface and the air the cold surface; however, as we have already seen, for a system employed outside of a pressurized zone, the air temperature will significantly decrease with the altitude, with the decrease in T_∞ at least partly compensating for the decrease in h;

– without it being systematic, during flight the relative wind will tend to create a forced convection, for example inside the wings, or engine pylons. Forced convection is highly favorable to convective exchanges[46];

– in cases of installation in motor zones, acoustic noise will also favor convection.

Figure 2.19 shows a test coupon for vibrations to which heat sinks of various geometries have been attached for validation purposes before test flights. In particular, this figure shows heat sinks with rectangular or cylindrical fins.

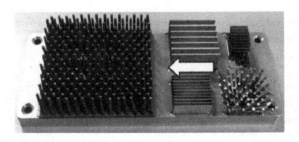

Figure 2.19. *Different kinds of heat sinks grouped together for a test. The metallic support is 35 cm in length. At the top right, there is a Micropelt GmbH module. For heat sinks with rectangular fins, the arrow indicates the ideal direction for air flow (authors' photograph)*

2.4.3. *Using vibrations*

In the process of establishing a list of ambient energy likely to be harvested in an aircraft in order to supply embedded wireless systems, vibratory energy will generally appear toward the top of that list. This form

46 However, it is important to be aware of new generations of aircraft for which aerodynamic structures are increasingly airtight and often require air ducts for cooling purposes in order to recreate the convection where it has disappeared (see Figure 1.3).

of energy is found on board an aircraft and brings together very different phenomena:

– random, low-frequency oscillations (<10 Hz) in the structure of the aircraft, caused by the flight (changes to the acceleration caused by the steering or by the engines, or by turbulence);

– permanent vibrations (and periodic at least by interval) in the structure of the aircraft, between a few dozen hertz and several kilohertz. These vibrations are caused by the engines;

– acoustic vibrations (noise), too weak inside the cabin to act as an energy source, but potentially much stronger where the engines are located, for example.

In the following section, we will discuss the nature of these phenomena and the transducers which could be used in order to transform this mechanical energy into electrical energy. We will leave the matter of aeroacoustics to Chapter 4.

2.4.3.1. *Low-frequency oscillations*

When an aircraft changes altitude or is subject to turbulence, accelerations take place on its structure which one might be tempted to try to harvest. As part of the European *chist-era* program, the authors of this volume are involved in the *SMARTER*[47] project whose goal is to retrieve these accelerations through the use of a woven, piezoelectric material known as Macro Fiber *Composite*[48] (MFC). This transducer (and actuator), MFC, subject to strain, is capable of generating electrical voltage and current, and can supply self-sufficient sensors monitoring key areas which are subject to these same strains (for example, wing roots) [DAN 13].

We will look at more detail at piezoelectricity[49] in the following section, but it should be noted here that the piezoelectric effect is the property of a material (ceramic, semiconductor, polymer, crystal, etc.) to bias itself electrically due to the effect of a stress, or inversely to generate a strain due to the effect of an applied voltage.

---

47 The authors' partners are the Universities of Barcelona and Exeter.
48 Marketed by *Smart Material Corp.*
49 The piezoelectric effect was discovered by Pierre and Jacques Curie in 1880.

As part of the *SMARTER* project, the aim is the instrumentation of composite parts, in the wing root, or in a fuel tank and therefore difficult to inspect visually. An MFC unit is recommended in order to both harvest the energy linked to strain (up to 10 Hz) and to move onto a measurement of these strains in order to monitor structural aging. Figure 2.20 shows one such unit attached to a coupon designed to undergo tests.

As far as the authors are concerned, this project is the only one attempting to retrieve vibratory energy at very low frequencies in an aeronautical context. The reader should understand that this is an exploratory project and that there is currently no *ready to use* transducer available for such a frequency range. In the following section, however, we will examine the current range of commercial equipment used for the harvesting of vibratory energy.

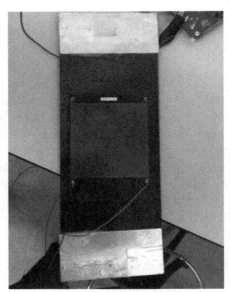

Figure 2.20. *IM6 carbon fiber panels to which an MFC-P2 piezoelectric transducer of 85 mm × 85 mm has been attached. This is then attached to an aluminum plate for stress tests (Photograph by Meiling Zhu, University of Exeter)*

2.4.3.2. *Harvesting of vibratory energy*

Theoretically, the harvesting of the mechanical energy from vibrations taking place onboard an aircraft seems an excellent idea, at least because of the distribution of these vibrations throughout the structure of the aircraft and the illusion of access to energy which is not dependent on the localization of the planned harvesting, as is the case in thermogeneration which requires the presence of a hot surface.

In reality, the situation is very different. Figure 2.21 shows a qualitative spectrum of vibrations measured close to the engine of a turbojet airliner with a capacity of more than 100[50]. For two of the three phases of the flight to be considered, this is a line spectrum: essentially, the frequency F_f of the fundamental is linked to the rotation speed of the blowing apparatus of the low pressure compressor, which generates periodic vibrations and therefore a line spectrum.

Unfortunately, the harvesting of this vibratory energy is made difficult due to the following characteristics:

– there is only energy available for harvesting at the peaks, but their position depends on the engine rpm, which necessitates a transducer with a sufficiently wide pass band (see below);

– when cruising, there are almost no power peaks, nor is there any energy to be retrieved, even though the cruising stage makes up the majority of the flight;

– for positions further away from the engine (for example, the fuselage), the spectral energy density is much weaker;

– for other positions on the aircraft, when a peak is present (for example, the fundamental), it can be shifted by a few Hz compared to F_f measured close to the engine (the filtering effect of the structure).

We will now examine the transducers that are available. For the most part, these come in three varieties [SPIE 15] (Figure 2.22):

– piezoelectric, which are in general (other than the structure known as MFC already presented) made up of a beam made (or covered) with a piezoelectric material, fixed at one extremity, and pulled by a seismic mass from the other, amplifying distortions and reducing the frequency resonance: due to the effect of the vibrations, an alternative potential difference will appear on the opposite sides of the beam.

---

50 For obvious confidentiality reasons, we are only able to publish qualitative results on this subject.

– electromagnetic, with an identical structure, with the only difference being that the beam is not piezoelectric but carries a permanent magnet (which also fulfills the role of the seismic mass) which produces a variable magnetic field in a fixed winding, and therefore leading to the appearance of an induced electromagnetic force (Faraday's law) in this winding (or inversely, the beam carrying the winding and the magnet is fixed).

– electrostatic (or electret[51]), which uses the variation in the relative position of the two conductive plates of an initially charged capacitor (or carrying a permanent charge in the case of an electret), and therefore produces a charge transfer if a closed electrical circuit is connected to it.

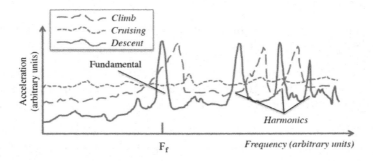

Figure 2.21. *Spectral power density of vibrations measured in flight close to a turbojet. The fundamental frequency F_f, identified here during descent is in the range 40–50 Hz. The vertical axis is qualitatively logarithmic and extends over roughly two decades*

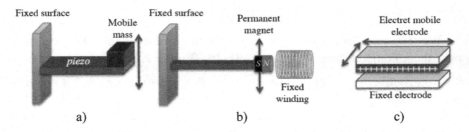

Figure 2.22. *Three mechanical energy transducers: a) piezoelectric (so-called cantilevered structure), b) electromagnetic (likewise) and c) electrostatic. The first two systems only enable unidimensional retrieval; for certain electrostatic systems, this can be bidimensional*

---

51 An electret is a dielectric material which carries a permanent electrical charge.

The kinds of technology involved in these three examples use MEMS-type technology for the first (partly) and the third (fully), while the second uses micromechanics. Generally speaking, miniaturization is the aim, which is only really possible for electrostatic systems. As a result, the power values delivered are less than 10 or so μW.

More problematically, the passbands of these systems, efficient when they are used at their frequency resonance, are only a few hertz (typically 3 Hz). Only large-scale commercial products[52] exceed 10 Hz for the passband, and are effective for accelerations (rather high) of 1 g R.M.S. If we are to consider that all of these transducers require moveable parts (for example, unlike a thermopile) and that the aim of aircraft manufacturers, for new generations of aircrafts, is to reduce the level of vibrations, the reader will understand that, in its current state, the authors of this volume have not implemented these systems on an aircraft.

For these three types of technology, commercial products do exist and correspond to certain types of application; below are the names of a few manufacturers:

– piezoelectric systems: *Midé Volture* and *Piezosystemjena*;

– electromagnetic systems: *Perpetuum Ltd*;

– electrostatic electric systems: Electret Energy Harvester Alliance.

2.4.3.3. *A few points regarding acoustic energy*

The question of how to appropriate it is to retrieve acoustic energy cannot be avoided in the aeronautical context: primarily because an aircraft is known to be a source of noise, and therefore a cause of the diffusion of sound waves in the environment, which theoretically makes their harvesting possible at a distance from the source of the emission.

To introduce the subject, it should be noted that the energy transported by a sound wave is characterized by the acoustic intensity I (in W/m^2) which is often expressed in decibels through its relationship with a reference acoustic intensity I_0 of 1 pW/m^2, through the equation:

$$L = 10.\log_{10}(I / I_0) \hspace{4cm} [2.11]$$

---

52 PMG17 (a mass of 650 g) from *Perpetuum*, for example. This system delivers 45 mW in the conditions given in the text.

where L is expressed in decibels, a unit generally abbreviated as dB_A (A for acoustic), dB_{SIL} (SIL for *Sound Intensity Level)* or finally dB_{SPL} (SPL for *Sound Pressure Level,* because in the case of a plane wave, acoustic pressure and intensity have identical values). The application of formula [2.11] makes it possible to make a connection between decibels and mechanical energy carried by the sound, which is for example:

- 100 dB_A corresponds to 1 $\mu W/cm^2$;

- 120 dB_A corresponds to 100 $\mu W/cm^2$;

- 140 dB_A corresponds to 10 mW/cm^2;

- 160 dB_A corresponds to 1 W/cm^2.

In order to clarify these concepts, the average sound environment at a concert should not exceed 105 dB_A (with a peak of 120 dB_A), while the sound of a pneumatic drill is roughly 120 dB_A. From the point of view of harvesting mechanical energy, if we are to consider the negative impact of the efficiency of the transducer, the harvested energy will therefore be modest at these levels.

In the immediate vicinity of a jet aircraft engine, depending on the engine model and the localization considered, the maximum acoustic level at take-off is somewhere between 150 and 160 dB_A. In this context, we have been able to prove in laboratory experiments that it was possible to supply an electronic system through acoustic/electrical transduction, for acoustic intensities greater than 120 dB_A (Figure 2.23).

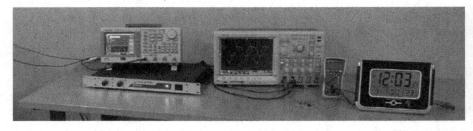

Figure 2.23. *Demonstration in a laboratory setting of energy supply by electrical/ acoustic transduction for a radio-controlled clock. The transducer (not visible) is a KPE-141 Kingstate piezoelectric buzzer subject to the acoustic intensity created by a loudspeaker (Sphynx SP-W65-SONO at 120 W) supplied by active equipment present in this photograph (author's photograph)*

However, for the following reasons, we have not pursued our studies in this direction.

– acoustic intensity rapidly decreases with distance from the engines, as well as during climb and while cruising (from roughly 10–20 dB$_A$ in this last case compared to take-off);

– aircraft manufacturers and motorists are constantly looking to reduce this intensity, and any energy harvesting mechanism runs the risk of not being durable.

Despite that, we have not given up on acoustic energy and we will demonstrate a concept relating to the retrieval of *aeroacoustic* noise in the last chapter of this volume.

2.5. Conclusion

In this chapter, we have introduced the three basic stages for a system of harvesting and managing energy designed for the autonomous supply of a system onboard an aircraft: transduction, storage and DC/DC regulation. We have in particular developed several processes for harvesting and transduction, as well as addressing the question of storage.

In Chapter 3, we will look in more detail at the electronic parts, preprocessing of the energy downstream from the storage, DC/DC convertor regulators and electronic managing structures.

3

Architectures and Electric Circuits

3.1. Introduction

The large majority of energetically autonomous embedded systems are powered by a battery or an accumulator. Most often, using a non-rechargeable battery is proven to be financially unbeatable. So, for a manufacturer working on wireless sensors, an application consuming a total energy of less than 10 Wh over its lifespan will preferably be powered by a battery. For a total energy consumption that is initially much higher, it would be better to minimize the electronic consumption rather than implementing an energy harvesting and storage system. However, batteries have limitations when conditions such as lifespan, cost of maintenance and usage are demanding. Using a rechargeable storage component (rechargeable accumulator and/or supercapacitor associated with the harvesting of ambient energy) is then required. For some applications, particularly with strong temperature restrictions as in aeronautics (–40, +80°C) or operating over a very long period, rechargeable accumulators could not be used and storage should be established around an electrostatic-type component (capacitor or supercapacitor).

The architecture of the "power supply" function of an autonomous embedded system with an ambient energy harvesting system is represented in Figure 3.1. Such a device obviously requires an energy resource in its surroundings, available permanently or intermittently. An adapted transducer linked to well-chosen electronics maximizes recovered electric energy under levels of voltage or current that are compatible with the following stages. When consumption coincides with production, the storage stage remains minimal and may be achieved simply with a capacitor that acts as a buffer to reduce inevitable differences between consumption and production or potential spikes in consumption. In other situations, storage will be achieved with one or several supercapacitors. Remember that this issue was broadly raised in Chapter 2. An adaptation stage is therefore required to bring the voltage to a level that is compatible with the load. Finally, the set is driven by an energy management system of varying complexity.

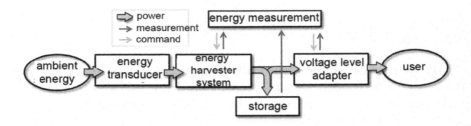

Figure 3.1. *Layout of the power supply function*

Let us now analyze the constraints associated with the absence of a battery. Compared to a system powered by a battery, the system could find itself without any available energy. It would then be good to initially provide a wake-up of the energy harvesting system when ambient energy returns, then, secondly, a successful load startup when the energy level in the storage stage is adequate. These two aspects are, respectively, treated in sections 3.3 and 3.4 of this chapter.

However, the huge voltage variability linked to the use of supercapacitors has an influence on the electronics of the energy harvesting system and on the voltage level adapter. The scaling of converters that they contain will be studied in section 3.5. And finally, a "safeguard" component will conclude this chapter with a presentation about electronic devices devoted to this usage.

3.2. Different storage modes

3.2.1. *System without storage*

Storage is only justified when the power required by the load is greater than the power delivered by the energy harvesting system associated with the embedded system. In some applications, there is no need for a formal storage stage, a small buffer capacitor is sufficient. This is the case for applications where a source of permanent ambient energy allows for a powered system to function continuously. Unfortunately, ambient energy harvesting uses sources that are often intermittent, and as it is unlikely for an operation rate that is more or less random to be accepted, so it is necessary to include storage.

We will make note that certain devices that use a passive telemetry [JAC 10] operate without storage, but with transmission distances of a few centimeters and limited possibilities of readings.

Passive RFID tags also function without a battery. Composed of a chip and an antenna, they use energy generated by an interrogator's magnetic or electromagnetic wave to supply power to the embedded electronic circuit and return information by using retro-modulation [BAR 05]. The range is a function of the device's frequency but does not exceed a few meters for this type of passive transponder without a battery.

3.2.2. *System with electrochemical storage*

If we wish to increase the lifespan of a system from an energetic point of view, it is easiest to replace the battery with a rechargeable accumulator and to connect it with an energy harvesting system to make sure that it recharges. However, electrochemical accumulators can only provide a limited number of charge/discharge cycles and their capacity diminishes over time. In this way, McCullagh [CUL 14] discovers, after 13 months of operation, a loss of data on a system that measures vibrations on a bridge, and the explanation for this is a drop in the accumulator's capacity.

It could be interesting to minimize in size and preserve the rechargeable accumulator by delegating cycles to a supercapacitor. The accumulator will function only as a backup when ambient energy harvesting defects. In this way, Alberola (Figure 3.2) uses a supercapacitor for everyday cycles with recharging done by a photo-voltaic panel, the lithium-ion battery does not supply power until after a period of bad weather.

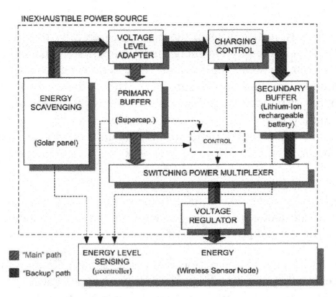

Figure 3.2. *Architecture of a system with the main power supply path and a backup path (according to [ALB 08])*

The system then benefits from the battery's energy store but works most often with the supercapacitor. Switching from one power supply to another is done based on the supercapacitor's voltage level. The accumulator's recharge strategy varies according to the authors [CAR 11] and [JIA 05]. Driving charge/discharge devices is digital or analog and it integrates the battery's temperature reading or not.

The autonomy of the device in absence of environmental resources is provided by the electrochemical accumulator, while the supercapacitor is scaled for regular cycles. The accumulators near constant voltage facilitate a possible reset if necessary.

3.2.3. *Storage made using supercapacitors*

It is possible to conceive of storage that is achieved solely by using supercapacitors. Two types of applications are targeted: aeronautical applications where batteries are often excluded (see Chapter 4) and applications with very low consumption (network of low consumption wireless sensors) where supercapacitors could replace rechargeable batteries without costing more and without a prohibitive increase in volume.

The design of the management part of an embedded energy system powered by supercapacitors is governed by performance (specific power and temperature stability) and defects (leakage current, variable voltage and maximum voltage) of these components. For illustration purposes, the following section compares a supercapacitor performance (Maxwell's PC10) and a lithium-ion accumulator (Multicomp's LIRR 2450), parts with similar thickness and mass (Figure 3.3) and analyzes the impact that these performances have on the scaling of the electronic energy management system.

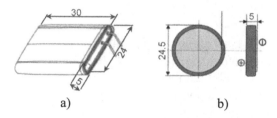

a)   b)

Figure 3.3. *Supercapacitor PC10: a) and accumulator LIR2450; b) the sides are in mm*

Supercapacitor PC10		Lithium-ion battery LIR2450
2.5	Voltage (V)	3.6
10	Capacity (F) / capacity (mAh)	120
6.3	Mass (g)	5.2
5	Thickness (mm)	5
2357	Volume (mm^3)	3,600
31	Energy (J)	1,555
1.4	Energy density (Wh/kg)	83
12.5	Max current (A)	0.24
–40, +70	Temperature (°C)	–20, +45 (charge) –20, +60 (discharge)

Table 3.1. *Comparing supercapacitor / rechargeable battery*

Remember that the energy of a battery or an accumulator is given by:

$$E = C.V \tag{3.1}$$

where E is the energy in Watt.hours (Wh), C is the capacity in ampere.hours (Ah) and V is the voltage in volts (V). For small quantities of energy, the joule is used: 1 Wh = 3,600 J.

For a supercapacitor, energy is given by:

$$E = \frac{1}{2} C.V^2 \tag{3.2}$$

where E is the energy in joules (J), C is the capacity in farads (F) and V is the voltage in volts (V).

3.2.3.1. *Specific energy*

Table 3.1 elucidates how weak supercapacitors are compared to accumulators: they have a specific energy that is much lower. For a system without a battery, we will then look to scale the energy reservoir without any excess, but this implies having good knowledge of energy budgets required by the load and ancillaries, as well as bringing in the energy harvesting stage.

3.2.3.2. *Specific power*

When it comes to power and therefore the current delivered or received, supercapacitors performed a lot better than batteries. Difficulties in providing current spikes, especially when starting the system, will not create a problem with supercapacitors, as long as the adaptor stage placed between the supercapacitors and the load can deliver the said current spike. When charging supercapacitors, limits to current are also not as strong as an accumulator that generally accepts a maximum load with a current of 1 C (i.e. a current of 120 mA for an accumulator with a capacity of 120 mAh).

3.2.3.3. *Voltage stability*

Storage component voltages vary according to the technologies used, but in addition, when in use, fluctuation ranges of the two families are very different. While the voltage in the LIR2450 accumulator given for 3.6 V varies between 4.2 V (end of charge) and 2.75 V (typical value for the end of discharge) with a voltage plateau of 3.6 V (see Figure 3.4), voltage of the PC10 superconductor varies between 0 V and its rated voltage (2.5 V for the PC10) with a value of 2.7 V which must not be exceeded so that the product is not damaged. Just like accumulators, one must therefore include a voltage control system at the storage stage in supercapacitors. When supercapacitors are assembled in series to increase their usable voltage, it is recommended, in order to prevent premature wear, to install a balancing system between supercapacitors. We will choose one with the lowest consumption possible. Experience has shown us that with 2 or 4 S (2 or 4 PC10 supercapacitors assembled in series), it is hence possible to do without balancing (by choosing supercapacitors from the same bundle and with very similar characteristics).

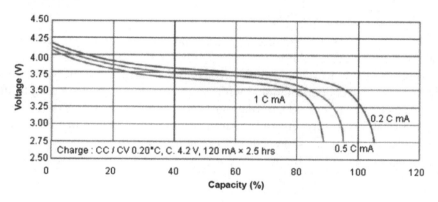

Figure 3.4. *Typical discharges in an LIR2450 accumulator with different currents (according to the technical sheet)*

3.2.3.4. *Cycle of life and aging*

While a PC10 supercapacitor can perform over 500,000 cycles, like all accumulators, the LIR2450 sees its capacity diminish over time to drop by 20% at 500 cycles (Figure 3.5). The possibility of a very large number of cycles constitutes a huge advantage for supercapacitors, the second advantage being the permitted temperature fluctuation as shown *infra*.

Discharge Characteristics

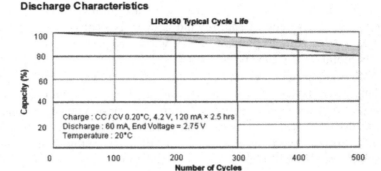

Figure 3.5. *The number of cycles influence on the capacity of an accumulator (according to its technical sheet)*

3.2.3.5. *Temperature influence*

Table 3.1 shows a range of uses for a supercapacitor that is much greater than uses for an accumulator. For aeronautical applications (–50, 80°C), it is no longer possible to use accumulators without a heating system while some supercapacitors remain functional with parameters that vary according to the model considered. For illustration purposes, Figures 3.7(a) and (b) infra shows variations in capacity and in series resistance measured in a PC10. For this analysis, the simple model shown in Figure 3.6 is used.

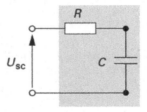

Figure 3.6. *Simple model of a supercapacitor*

Readings were carried out with charges and discharges with a constant current of 300 mA. We will make note of a usage area broader than the one given by the manufacturer (–50, +100°C) and with relatively low and, most importantly, reversible parameter variations.

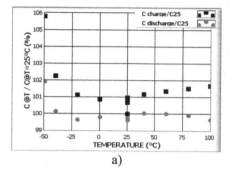 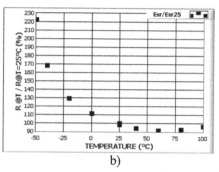

a) b)

Figure 3.7. *Relative variations in capacity a) series resistance and b) as a function of temperature for a PC10 HT supercapacitor (readings carried out by the authors, index 100 at 25°C)*

3.2.3.6. *Leakage current and auto-discharge*

A simple model composed of a capacitor C and a series resistance (ESR) is generally enough to describe the behavior of a supercapacitor. However, for embedded systems brought to operate most often in a very low consumption mode[1] (*sleep* mode with a typical consumption of between 1 and 10 µA under 1.8–3.3 V) and more rarely in active mode, it is good to be aware of the auto-discharge in the storage device. In fact, except for a few models, most supercapacitors have a large auto-discharge. To model these phenomena, a leakage resistance is added in parallel (Rleak) connected to branches with high time constants. The initial simple diagram is hence completed as shown in Figure 3.8 [MER 12], the figure in which the nonlinear behavior of supercapacitors begins to appear.

Figure 3.9 shows the practical impact of the phenomenon. At the end of 3 days, with a Maxwell-type PC10, for example, auto-discharge leads to a drop in voltage by 6%, which corresponds to an 11% loss in stored energy. For other models, the drop in voltage is much more significant and we must remember that when voltage drops by half, energy is divided by 4. It is important to choose the supercapacitors used carefully.

---

1 In other words, for a source of intermittent ambient energy and in a long-term operation relying uniquely on internal storage.

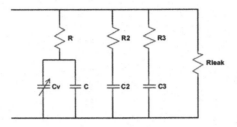

Figure 3.8. *Electric model equivalent to a supercapacitor. We note the main R C branch (C is the supercapacitor usual capacity), the Cv branch, Cv being dependent on V, "slow" branches R2C2 and R3C3 and finally resistance Rleak modeling leaks (auto-discharge)*

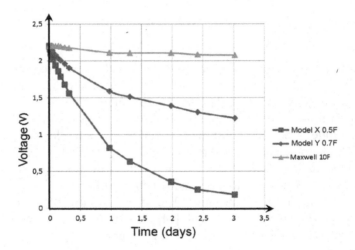

Figure 3.9. *Voltage progression at the terminals of three supercapacitors*

Modeling the phenomenon is complex [WED 11] and difficult to use in order to carry out a scaling of a supercapacitor. Empirically, starting with the auto-discharge curve carried out previously, we can evaluate the power lost during auto-discharge by considering equation [3.2] and the formula:

$$P_f = \frac{\Delta E}{\Delta t} \qquad [3.3]$$

where P_f is the supercapacitor's leakage power.

Figures 3.10 and 3.11 show the progression of leakage power for the same supercapacitors during the auto-discharge phase, respectively, as a function of time and voltage at the terminals of each supercapacitor.

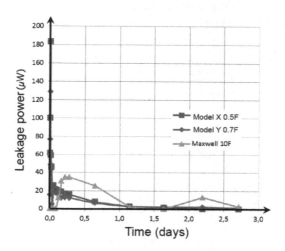

Figure 3.10. *Leakage power as a function of time*

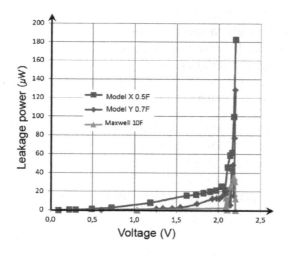

Figure 3.11. *Leakage power as a function of voltage at the superconductor's terminals*

We see in the first half hour (the very start of Figure 3.10) different behaviors according to the models. The large leakage power at the beginning of the experiment for two of the three models is explained by the influence that the slow branch of the supercapacitors has: we see a recombination of electric charges in the fast branch toward the slower branches. For the PC10 supercapacitor, this phase is missing because the PC10 remained for quite some time under a constant voltage before the beginning of measurement[2], which is not the case for other models. After this first phase, we read a minimum leakage power that decreases with the residual voltage.

By being acutely aware of these phenomena, a complex modeling would be required, but it is possible to use an upper limit in the second phase (for example, a 40 μA one here for the PC 10 model) and eventually rely on the voltage of supercapacitors to not overestimate the phenomenon.

3.2.3.7. Conclusion

Generally, the consumption required by the load is well known. The difficulty with scaling the storage stage in aeronautical applications results in constraints expressed above that we could summarize as follows: a potentially unpredictable resource according to the proposed scenarios and significant variations in component parameters (a link with variations in temperature). This leads to carrying out either a scaling in the worst case with a risk of over-scaling the storage stage and the associated volume, or to accept an operational limitation from the outset, or an intermittence, or operating in downgraded mode. In all situations, this leads to a reduction in service quality, which is generally unacceptable.

Once the initial energy measurement is established and the base element for storage is chosen, the layout of storage components (placed in series and placed in parallel) is carried out as a function of possible choices (maximum levels of current and voltage). Taking into account various errors (ESR, effect of temperature, etc.) allows us to refine in a second phase, the initial scaling.

---

2 The auto-discharge dynamic is basically a function of the way in which the supercapacitor was charged.

3.3. Set up and operation of the energy harvesting system

3.3.1. *Initial startup*

During set up, the system could be initially without any stored energy, and therefore without any control. Despite everything, when the energy resource appears, the energy harvesting system must naturally begin operation. Most often, it starts up in downgraded mode. This is the case with commercial circuits dedicated to an energy harvesting system composed of a step-up layout type (*Boost*) like the LT3105 from Linear, the BQ25504 from Texas Instrument or the SPV1040 from STMicroElectronics. You must therefore wait for enough voltage in the storage stage so that the converter goes into Boost mode searching for the maximum power point (command (MPPT). We therefore observe for these circuits a first phase for charging that is not very powerful, as the integrated control circuit cannot force the transducer to work alongside the maximum power point.

Interestingly, to minimize this phase during the following startups, it is worth maintaining a minimum energy reserve level when the surrounding resource is no longer there by cutting off the load's power supply before supercapacitors completely discharge. The voltage level of the storage stage must allow for a quick restart as soon as the external energy resource will be present again. This aspect will be studied in the next section.

An example of a system initially starting up in the "downgraded" mode, then switching on a more energetically powerful layout is shown by Boisseau [BOI 13a]. In this case, the first passive layout to charge a capacitor Cs (because the switch dMOS is initially normally conducting at start-up) is used in the startup of the energy harvesting system (denote by *EH* for *Energy Harvester*). The path of the current is marked in green in Figure 3.12. Once this capacitor is sufficiently charged, a *Startup control* unit (carried out with a Schmitt trigger) blocks the initial layout and authorizes powering the Control circuit which drives a more efficient Flyback-type layout and then charges the Cb capacitor (path underlined in orange). Capacitor Cs is recharged by Cb through the diode Dc. When the voltage at the Cb terminals is sufficient, the *WSN Control* unit then enables power to be supplied to the wireless sensor network (*WSN*). This lets it operate as soon as 10 µW of power is recovered.

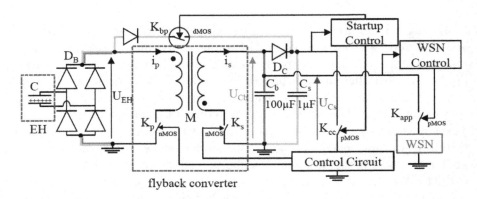

Figure 3.12. *Harvesting circuit and power management with an auto-power supply (according to [BOI 13a])*

3.3.2. *Startup of the energy harvesting system under low voltage*

When ambient energy is available, electric voltage is often initially small and less than the required level to startup a converter. Typically, powerful step-up converters begin to operate for input voltages of around 0.7 V. Once started up, they continue to operate until input levels are 0.3 V. It is important to supply an auxiliary source of voltage to supply power to the control circuits and to initiate startup. In this way, Spies [SPI 08] suggests using a charge pump powered directly by the energy harvesting system to startup a harvesting system using a thermal generator. The charge pump output powers the DC/DC converter (via dedicated inputs for certain integrated circuits – pin Vin or pin Batt on Figure 3.14 – or with pin OUT for the majority of these products). Once the DC/DC converter starts up, the charge pump is turned off or short-circuited.

Figure 3.13 illustrates the concept used, and Figure 3.14 shows the practical implementation where the charge pump short-circuited after startup because the step-up converter's control circuit gets power from its output *via* the D_2 diode.

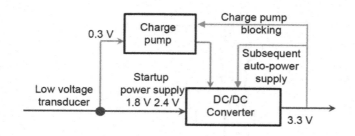

Figure 3.13. *Startup concept of a DC/DC converter (according to [BEC 08])*

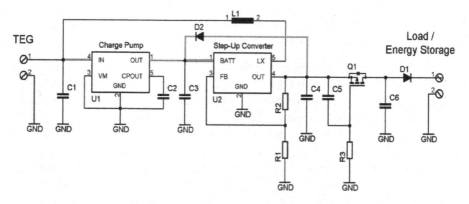

Figure 3.14. *Example of a layout using a charge pump to initiate startup (according to [SP I 08])*

3.3.3. *Operating the energy harvesting system under low voltage*

We have just demonstrated the use of a branch circuit (charge pump) to help startup the main circuit. However, when voltage levels at the transducer output remain at very low levels, below 0.3 V, it is better to use other topologies. The natural idea is then to use a step-up transformer; but in order for it to operate correctly, an alternate voltage is required. With thermal generators or solar cells that supply DC voltage, one must consequently set up an oscillator. The basic layout is shown in Figure 3.15.

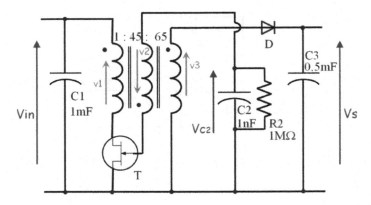

Figure 3.15. *Concept diagram of the step-up assembly (according to [DAM 97])*

With a step-up transformer with three windings, the key component in the circuit is the transistor JFET (T), normally on. If a small positive voltage appears at the input Vin of the converter, a current passes through the transformer's primary winding. This current increases exponentially over time and as the voltage V_1 is positive, you then find a positive voltage V_2 at the terminals of the secondary winding. The terminal coming out of this secondary winding is connected to the T_1 grid being maintained at a constant potential by the intrinsic diode of JFET T, the other terminal of the secondary winding is connected to a negative potential, which ends up negatively charging C2 with V_{C2}. When the current at the primary saturates the magnetic core, the variation in current and therefore the induced voltage in the secondary cancel out, which produces a fall in the secondary voltage V_2. The sum of voltages in the capacitor C_2 and the transformer's secondary then becomes negative, which blocks T. Since the desired voltage at output must be positive, we cannot use the voltage V_{C2}, the third winding then lets us have a positive voltage that is enough to charge C3 through the diode D. C2 discharges through R_2. The JFET becomes conductive again and the cycle can be repeated.

A more elaborate version is used on Linear Technology's LTC 3108. The concept diagram for the input stage of the circuit is shown in Figure 3.16. We see the same idea for the oscillator, but with only one secondary. This circuit operates from 20 mV and provides an adjustable voltage between 2 and 5 V at output.

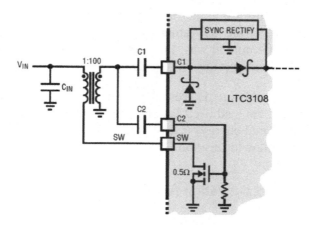

Figure 3.16. *Input stage for LTC3108 (according to its technical specifications)*

In the example for the application of the circuit shown in Figure 3.17, we see different capacitors; a high capacitor of 0.1 F capacity in charge of storage (Vstore) and two others with much lower capacity values lead to a rapid startup and allow the circuit to become operational very quickly.

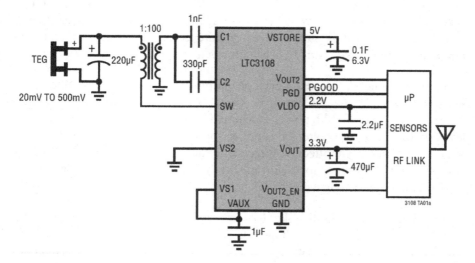

Figure 3.17. *Wireless sensor powered by a thermal generator (extract of the technical documents for LTC3108)*

At startup, it is useful, in the first instance, to quickly charge a low value capacitor to allow for the harvesting circuit to go into power mode. Then, in a continuous or intermittent way, the output of the harvesting system will be connected to the storage stage.

In this section, we are interested in starting up the energy harvesting system; generally, the load itself is only activated with a delay, when enough voltage (and/or stored energy) is available. The role of the UVLO circuit is looked at in the following section.

3.4. Delayed load activation (undervoltage lockout – UVLO)

3.4.1. *Illustration of problems*

With a system powered by a battery, load startup is instantaneous once the battery is connected, on the condition that the battery's voltage is in compliance with the load's operational range and that this supply voltage fluctuates very little with the battery's charge status.

However, the voltage V_{SC} at the terminals of a supercapacitor depends on the level of charge. If we think about the classic layout made up of two supercapacitors in series, the level of voltage typically varies between 0 and 5 V, while the load is supposed to operate correctly with an operational range made up of V_{MIN} and V_{MAX} which has its implications.

First, in general, the device's consumption is zero or negligible for a voltage that is below the level V_0 (with $V_0 < V_{MIN}$). When the voltage achieves this level V_0, the power consumed by the load increases to a certain level. If this level is greater or equal to the recovered power in the energy harvesting stage, supplied voltage will never reach V_{MIN}, a required condition for the proper load startup. So the load must not be connected at the storage stage as long as $V_{SC} < V_{MIN}$.

In addition, a current spike I_{PIC} is often present during startup. This current spike creates a voltage drop V_{DROP} at the terminals of the supercapacitors because of the series resistance in the supercapacitors ($V_{DROP}= ESR.I_{PIC}$). Since this takes place with a supply voltage close to V_{MIN}, the system will shut down immediately. To avoid such a situation load startup must be delayed.

Finally, when the recovered power is in short supply, the system is powered using an energy reserve in which the voltage gradually decreases. When it falls below the threshold V_{MIN}, the load no longer operates correctly but its consumption will remain stable or increase, which will end up emptying supercapacitors and delaying the next startup. It is therefore important to cut the power supply as soon as the voltage falls below V_{MIN}.

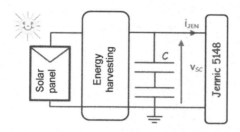

Figure 3.18. *Unit-diagram of the test device and readings carried out*

To illustrate these statements, we will give an actual example of readings carried out by [BOI 13b] on the system described in Figure 3.18. Different elements of the system are as follows:

– storage is made up of two superconductors (Maxwell's PC 10) in series;

– the power supply is provided by solar cells (MSX – 01 panel V_{OC} = 10 V, I_{SC} = 150 mA @ STC[3]);

– an optimized energy harvesting system based on a Buck converter layout and driven by a microcontroller was fitted between the panel and the storage stage [MEE 10];

– the load is a *JN5148* module targeted for low-power applications adapted for wireless networks. Here, it is programmed to carry out a reading and to transmit every 8 s and to remain at rest between two readings. The unit's specifications show the correct operation for a supply voltage of between 2.3 and 3.6 V.

---

3 Standard Test Conditions: 1,000 W/m² (optical) at 25°C.

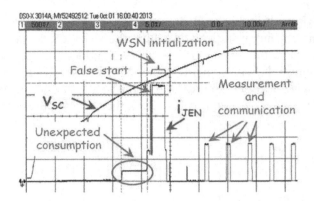

Figure 3.19. *Initial startup with significant recovered power*

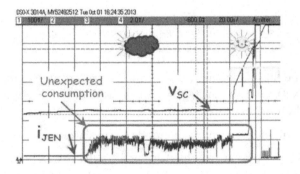

Figure 3.20. *Initial startup with low recovered power in the primary phase; initial V_{SC} =2.0 V*

Figure 3.19 shows the increase in voltage for supercapacitors and current supplying power to the unit when the initial discharged device is placed under constant irradiation (good sunlight). At startup, voltage V_{SC} increases steadily and the unit does not consume, which is normal. When the voltage V_{SC} reaches 2.0 V, consumption becomes equal to 2 mA and stays at this value until V_{SC} reaches the unit's low threshold startup of 2.3 V.

We must also note several false starts after the phase at 2 mA (significant oscillation of the startup current at initialization which translates into malfunction): the initial power spike remains difficult to sustain as long as the voltage of supercapacitors remains just enough.

In the example in Figure 3.19, the unit is initialized because the current generated is greater than 2 mA. However, in Figure 3.20, carried out with much lower irradiation and therefore lower recovered power, the device remains blocked in this unwanted phase and does not start up again as long as the resource does not increase.

In conclusion: at startup, if the recovered power is not greater than the power consumed by the device that must be supplied, the system can obviously not startup immediately, but on top of it, it uselessly consumes the low amount of energy already stored if startup was not delayed.

A UVLO-type device lets you delay and/or cut the power supply to the load and solves the problems described in this section. It is made up of two elements:

– a pilot switch that allows or does not allow power to be supplied to the load;

– a control logic.

The following sections describe different possible solutions for carrying out these two subsystems.

3.4.2. *Carrying out the UVLO*

3.4.2.1. *Switch on the mass line*

This solution is simple to implement (see example in Figures 3.21 and 3.22). An NMOS acts as a power switch. Its voltage V_{GS} is driven by the control logic. For the substrate diode (called body, intrinsic diode of the MOS) to never be conducting, an NMOS must be used.

Resistance in parallel between the grid and the source maintains a zero voltage in the absence of grid polarization, while this guarantees an open switch (*Switch Off*) during initial startup (*cold start*). A high value for resistance limits losses when the control signal (V_{BAT_OK}) is at a high level.

The choice of NMOS is dictated by the following considerations:

– a threshold V_{GS} (*Gate threshold Voltage*) compatible with voltages in the control logic;

– a lowest possible MOS resistance in the conducting state $R_{DS(on)}$[4] to minimize losses in the MOS in the conducting state and to maintain the drop in voltage $V_{DS} = R_{DS(on)} \cdot I_{LOAD}$ at an acceptable level ($V_{LOAD} = V_{SC} - V_{DS}$), a very high drop in voltage could deactivate the load;

– the MOS must support the current required by the load, particularly the startup current that could be high on a capacitive load.

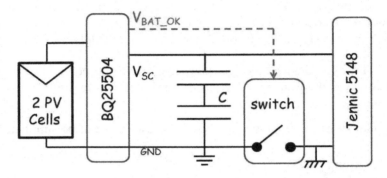

Figure 3.21. *Example of the layout with a switch on the mass line*

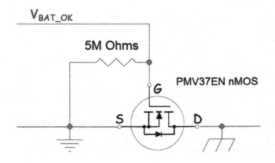

Figure 3.22. *Carrying out a switch on the mass line*

This layout's disadvantage: when the switch is open, potentials to the left and right of MOS are different. Especially in the test phase, one must pay

---

4 In a MOS, $R_{DS(on)}$ is the total resistance between source and drain during the on state determining loss when conducting a current.

attention to this loss of continuity in the system's mass. To minimize this disadvantage, the following layout could be used.

3.4.2.2. *Switch on the power supply line*

Figure 3.23 is an experimental assembly using a switch on the upper line. Switch details are shown in Figure 3.24, the control logic is shown in Figure 3.25. With a layout such as this, there is only one mass in the assembly, which simplifies taking readings in the test phase.

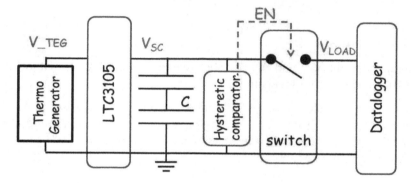

Figure 3.23. *Assembly with a switch on the power supply line*

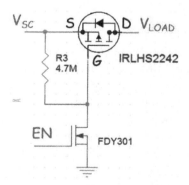

Figure 3.24. *Discrete switch implementation*

On the power supply line, the switch is made up of a transistor MOSFET channel P or PMOS where the grid is driven by an NMOS. The intrinsic diode of the MOS requires taking a PMOS.

A "high" logic control makes the PMOS conductive by setting its grid potential to 0 V via the NMOS which is conductive. Inversely, a "low" logic control level opens the switch by blocking the NMOS, the potential of the PMOS grid being brought to the V_{SC} level by the intermediary of the pull-up resistance.

The important parameters to consider are:

– resistance $R_{DS(on)}$ of the PMOS, with the same considerations as seen previously for the NMOS;

– input levels V_{IH} and V_{IL} of the switch's control unit that must be compatible with output levels in the control unit;

– the switch's slew-rate which limits current when closing the switch (*inrush current*). We sometimes add a capacitor between the drain and grid of the PMOS and a resistance in series with the NMOS [PIC 08] to modify the slew-rate of the PMOS.

Such a switch is easily achieved with discrete elements, but it also exists in an integrated form as shown below.

3.4.2.3. Load switch

A load switch includes a PMOS that operates on an input range (for example, between 1 V d 3.6 V for a TPS22901). The switch is driven by a logic signal that is compatible with classic control levels. As an option, a resistance is added in parallel at the output to avoid leaving the load at a floating potential and to make the load's voltage drop quickly when the switch opens. Other models like the TPS 2100 (Vaux Power-Distribution switches) could also be productively used. A very low consumption in these circuits makes them compatible for low-power applications (current consumed less than 1 µA in open switch mode and 7 µA under 1.2 V in conduction mode for the SiP32411 from Vishay Siliconix).

The load switch provides:

– a low $R_{DS(on)}$;

– a very compact design;

– a power supply current <1 µA;

– a controlled slew rate;

– the guarantee that the output potential is not floating but remains at 0 V when the switch is open, as the concept diagram taken from SiP32411 specifications demonstrates it well (Figure 3.25).

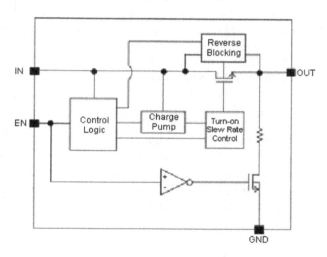

Figure 3.25. *Concept diagram for a load switch according to the technical specifications of the SiP32411 from Vishay Siliconix*

This solution is very useful. However, an even easier solution is sometimes available as shown infra.

3.4.2.4. *Using a "Pin Enable" -type control*

Devices to be supplied with power sometimes have a *Pin Enable* or are powered through a regulator that has this kind of input. In this case, the solution using this control is the easiest to implement. It is useful for verifying that activation and deactivation levels are compatible with the control logic. It is also useful for verifying that the induced consumption, when the *Enable* input is at a low state (quiescent and shutdown current), remains low. In the opposite case, we return to previous solutions.

3.4.3. *Control logic*

Let us look now at feasible solutions for the implementation of control logic.

This system must read V_{SC}, and as a function of its measured value, generate a logic signal to connect or disconnect the device that is to be supplied power. If a microcontroller is in the system, it could carry out this task as long as it recovers the voltage V_{SC} on an analog to digital conversion input and to dedicate a digital output to this function.

In analog mode (without a microcontroller), a hysteresis comparator with two thresholds V_{IH} and V_{IL} could carry out the required function. As an UVLO system has to block the power supply of a device as soon as it can no longer operate correctly, we will then adjust V_{IL} so that $V_{IL} = V_{MIN}$.

For applications powered by supercapacitors, the control logic must delay the set up or more specifically wait to validate activation, as the supplied voltage V_{SC} must be greater than the required minimum V_{MIN} with a security threshold ΔV. In fact, the energy needed at system startup leads to a decrease in the voltage V_{SC} that must not fall below V_{MIN} at the end of startup.

However, the startup current conducted in the resistance of the supercapacitor's (ESR) leads to a drop in instantaneous voltage at supercapacitor terminals. In spite of this drop, V_{SC} must always be greater than V_{MIN}. It is useful to then take a high threshold V_{IH} that at least respects these constraints.

3.4.3.1. *Hysteresis comparator with discreet elements*

The control circuit in Figure 3.26 is implemented with discreet elements around an *ultralowpower* comparator. This comparator includes a voltage reference and power is supplied directly by voltage from the storage stage.

Passive components are chosen to minimize losses. This circuit consumes about 10 µA. It is used in the type of circuit shown in Figure 3.23. The experimental curves during startup are shown in Figure 3.27.

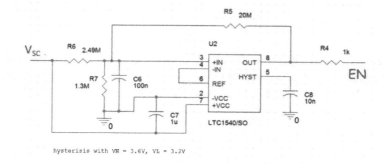

hysterisis with VH = 3.6V, VL = 3.2V

Figure 3.26. *Hysteresis comparator*

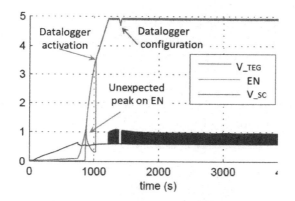

Figure 3.27. *Startup phase [MON 14a]*

With the power supply being provided by initially discharged supercapacitors, V_{SC} is initially zero. The comparator will be activated from a certain level of supplied voltage, with an unwanted phase: an unwanted spike on the control signal EN when V_{SC} reaches 1V. Luckily, the voltage level of this spike is too weak to activate the data logger (level is less than the V_{GS} threshold) and the spike has no effect, but this emphasizes possible problems. Here, a wise choice of elements avoids malfunction.

3.4.3.2. *Integrated hysteresis comparator*

The desired function is often directly implemented in the commercial energy harvesting circuit. As such, on the BQ25504, a V_{BAT_OK} signal is available. Three resistances easily set high (V_{IH}) and low (V_{IL}) switching thresholds.

When the solar resource disappears, the integrated circuit is fed from supercapacitors, and a consumption of between 6 and 7 µA was measured indirectly by removing the load and by visualizing the gradual drop in voltage at the storage capacitor's terminals (4,700 µF for this trial).

3.4.3.3. Dedicated integrated circuit

We can also directly use a dedicated activation circuit such as voltage detectors from the series XC61C by Torex or the MAX6375 from MAXIM. For these components, the high level (or low according to the model chosen) at output is guaranteed for a voltage greater than a threshold directly linked to the chosen reference and very weakly dependent on the temperature. However, passing through an unwanted state is inevitable when voltage increases at the component's input (voltage of the storage stage for our applications) as we see in the portion highlighted in the static characteristic of a XC61 voltage detector (Figure 3.28). Also, during system design, we must wait to have a trigger level greater than around 0.6 V for the next stage, or there will be an untimely trigger when Vsc < 0.6 V.

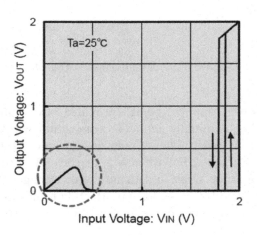

Figure 3.28. *Static characteristic of XC61. The threshold is at 1.8 V (according to the component's technical notice)*

3.5. DC/DC converters

3.5.1. *Functions*

DC/DC converters that we generally find at the energy converter level and the level adaptor are several different types: step-down (*Buck*), step-up (*Boost*) or step-down/step-up (*Buck-Boost*). We could encounter topologies with transformers (*Flyback and Forward*), which allow for galvanic isolation and, if required, permit to lower the voltage at the same time.

At the required functionality level, the DC/DC level adapter placed between the storage stage and the load should supply a regulated voltage level as a function of a load specification set by the load that needs to be supplied power, and it should be relatively insensitive to variations in the load current (*Load Regulation*) as well as variations in voltage at the storage level (*Line Regulation*). It should cut when supercapacitor voltage falls under a given threshold and is, if possible, protected against short-circuits at output (which is not normal for a Boost). These protection aspects will be dealt with more precisely *infra*.

Activation of this stage (*Startup*) will be managed autonomously or by a supervisor stage depending on the application.

The DC/DC associated with the energy harvesting system has different functionalities. It must optimize the harvesting of electrical energy and control the charge status of supercapacitors by potentially limiting the charge current. First, this involves search functions for maximum power (control MPPT *Maximum Power Point Tracking*) and second, measuring voltage at the storage stage to stay in the authorized range while trying to maintain an energy reserve at maximum capacity when it is possible to do so.

We then understand that the same converter will be lead over time to go from one mode to another depending on the phases considered. Digitally controlling the different modes is more convenient once specifications become complex. In low power commercial harvesting, circuits operate autonomously and directly integrate most of the required functionalities (startup, maximizing the energy harvesting, controlling the maximum voltage in the storage stage and cutting the charge in the event of low storage voltage under a given threshold). For more significant power (between 5 and

10 W), we have been led to develop specific converters, controlled by a digital special energy management system (microcontroller).

3.5.2. *Topology and scaling rules*

The topologies used (step-down, step-up, with or without isolation) are directly linked to voltage and current levels at input and output of each stage and the need or not of a galvanic isolation between stages.

Contrary to systems connected to accumulators, the extremely variable voltage range in supercapacitors is a serious constraint that must be taken into account when designing the circuit. So, for example, in the situation with a *Buck* positioned between a photovoltaic panel and an electrostatic storage stage that is initially empty, at startup the voltage V_{SC} is very low[5], as well as the duty cycle, which will induce a very high current (the *Buck* is a voltage step-down and therefore a current step-up). Luckily, this current is limited by the parasitic elements of passive components in the absence of an active control. This phase prolongs the time that the voltage V_{SC} reaches a correct level (at a constant input power, the input current in the storage stage decreases as its voltage increases).

If scaling components is only based on the rated plan (steady state), you can expect a few "surprises" during the transitional phases (startup or end of operation). For example, in an application, we placed a Boost-type step-up converter at the storage stage outputs (8 V of rated voltage) to supply power to the load under 40 V. This circuit, scaled initially for a rated operation, was not suitable when the supercapacitors finished emptying (overheating of the MOS of the Boost). Basically, thermal constraints became much stronger for this MOS for decreasing V_{SC} voltage because the duty cycle increased, and therefore the conduction time of the MOS. Being aware of operation worst-case scenarios consequently must also be the rule.

For embedded applications, volume limitations, even the available thickness, are also significant constraints. When system integration must be pushed to its limits, choosing inductance is often difficult. With current ripples being inversely proportional to the inductance value and switching frequency, we will therefore try to increase the switching frequencies in

5 The supercapacitor will behave much longer in a short-circuit depending on how high the value of its capacity will be.

conversion stages to minimize induction sizes. In return, a high switching frequency results in significant change-over switching losses. A size/performance compromise specific to the targeted application will be obtained through a rational choice in all system components to minimize losses.

Below, we take a look at two layouts that we had a chance to use for different studies: a *Buck* layout and a *Boost* layout. Many other layouts are possible, but the main useful ideas for scaling overall remain the same.

3.5.3. *Step-down converter*

3.5.3.1. *Layout*

The layout shown in Figure 3.29 is the converter layout. In this figure, what is not mentioned is the management circuit of the PWM (*MOS Driver*) that controls the high (T_H) and low (T_L) MOS as well as reading circuits required to make sure the operation of the converter is correct.

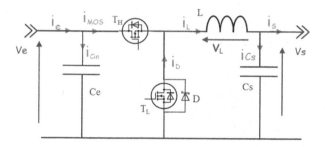

Figure 3.29. *Main diagram of a Buck operating in synchronous rectification*

We will definitely try to minimize device losses and the size of the circuit, which leads to switching frequencies between 100 kHz for powers that are 10 watts, to a few MHz for the smallest converters used for power with a magnitude of 1 watt.

Note that T and f are the switching period and frequency. The duty cycle is denoted by α. Transistor TH is conducting and transistor TL is open for time interval $[0; \alpha T]$; TH is open, TL is conducting for time interval $[\alpha T; T]$.

3.5.3.2. *Scaling of the input capacitor*

The input capacitor is chosen in a way to minimize ripples in the input voltage, as too much of a ripple is actually harmful to the energy harvesting system, particularly when the power recovered curve versus input voltage $P_{RECUP}(Ve)$ is quite a localized maximum, which is particularly the case with powerful solar panels.

We assume in the first case that the value for Ce is quite high so that Ve remains practically constant over a switching period. In this situation, the current provided by the harvesting system is also constant because the harvesting system works at the same point of operation. The current in the capacitor Ce is given by the relationship:

$$i_{Ce} = i_e - i_T \tag{3.4}$$

By taking the simplified hypothesis that the current in inductance remains constant and is equal to $<I_L>$, we get wave shapes as shown in Figure 3.30. The system, assumed to be in steady state at the end of the period, the voltage V_{Ce} goes back to its start value. The voltage V_{Ce} being an integral of the current i_{Ce}, positive and negative (hashed) areas are equal over a period, to obtain by integration $V_{Ce}(T) = V_{Ce}(0)$.

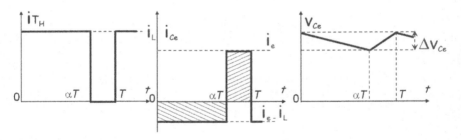

Figure 3.30. *Wave shapes obtained by ignoring inductance ripples*

Between αT and T, $i_{Ce} = i_e$ because $i_{TH} = 0$ (T_H is open). In this interval, voltage in the capacitor increases linearly with the slope I_e/Ce; we then deduce the ripple ΔV_e of the input voltage:

$$\Delta V_e = \Delta V_{Ce} = \frac{I_e}{Ce}.(1-\alpha).\frac{1}{f} \tag{3.5}$$

We generally hope for a maximum ripple at inputs of 5% of the maximum input voltage so that the input can operate at a stable operating point, which sets ΔV_e.

The maximum input current is given by possibilities with the energy harvesting system, or is limited by the *Buck* control at the desired value. We will use this value for maximum current for the calculation. The maximum value of $(1-\alpha)$ being 1, we deduce the minimum value for Ce:

$$Ce = \frac{I_e}{\Delta V_e} \cdot \frac{1}{f}$$

[3.6]

A higher value will minimize ripples. The calculation is more complicated if we consider the ripple of the current in inductance. A security factor 2 on Ce will be suitable without redoing calculations and a simulation will help to validate and/or refine initial choices if necessary.

3.5.3.3. *Scaling inductance*

The size of inductance will be minimized if the system is at its continuous conduction limit (Figure 3.31). We assume that this condition was achieved and we consider that input and output voltages remain constant in a switching cycle.

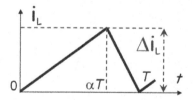

Figure 3.31. *Appearance of the current in inductance at the boundary of continuous/discontinuous conduction*

We then have a current increasing linearly from [0 to αT] and linearly decreasing at the end of the period, from αT to T.

Rippling of the current in inductance is therefore given by:

$$\Delta I_L = \frac{V_e}{L} \cdot T . \alpha . (1 - \alpha)$$

[3.7]

which is a maximum for $\alpha = 1/2$, with:

$$L = \frac{V_e}{\Delta I_L} \cdot \frac{1}{4f}$$ [3.8]

3.5.3.4. Scaling of the output capacitor

We first assume that the value for Cs is quite high so that the output voltage V_s can be taken as constant over a switching period. The output current i_s is also constant.

As the system is in steady state, there will not be an increase in the value for the voltage at the terminals for Cs over a cycle, and therefore $\langle i_{Cs} \rangle = 0$. The positive hash surface is then equal to the negative hashed surface (Figure 3.32). The average value for I_L goes to the load, the alternative portion in Cs. Aware of the form of the current i_{Cs} (portions on the line), we show through integration that the shape of the wave for V_{Cs} is made up of pieces of parabola. The calculation shows that the ripple on Vs is independent of α and is expressed in the form:

$$\Delta V_s = \frac{\Delta I_L}{8.Cs.f}$$ [3.9]

which gives:

$$Cs = \frac{\Delta I_L}{8.\Delta V_s.f}$$ [3.10]

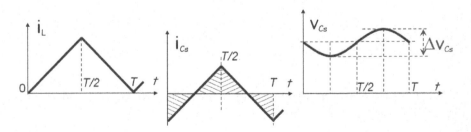

Figure 3.32. *Waveforms in the inductance and the output capacitor for $\alpha = 1/2$*

3.5.3.5. *Choosing of other devices*

Losses are distributed in passive components, in drivers and reading components, and in switching components.

So, it will be good to choose passive components with the smallest possible resistance in series at the desired switching frequency.

For switching devices, a very detailed analysis of losses during conduction and in switching is given in an application note by *Fairchild Semiconductor* [KLE 14]. Analysis of these losses shows in particular that losses in the high MOS (T_H component) are essentially losses in switching, while losses from conduction prevail in the low MOS (T_L component). So, one must choose MOS with a $R_{DS(on)}$ that is as small as possible to minimize losses in conduction and a charge $Q_{G(SW)}$ that is as small as possible to minimize losses through switching. One must also arbitrate between a low $R_{DS(on)}$ and the size of the MOS, the two parameters that grow in opposite directions.

The power components operating in switching are chosen to be able to sustain the required currents and voltages (a safety margin of 20% is used for the current and 50% for the voltage V_{DS} for MOS).

At the same time, carrying out the command requires choosing MOS's whose threshold voltage V_{GSth} is compatible with (and therefore less than) voltages delivered by the MOS driver. We make sure to choose drivers that are capable of delivering an adequate current so as to not slow changeover switching.

When blocking the high MOS, the low MOS becomes conducting but with a time delay that is managed by the MOS driver. This dead time is required to avoid a short-circuit on the bridge arm made of the two MOSs. During this transient phase when the low MOS is still not conducting, the role of the classical freewheeling diode is carried out by the intrinsic diode of the low MOS, until full conduction of this MOS. A Schottky diode, connected in parallel on the low MOS, chosen with a low threshold (typically 0.4 V@1A) less than that for the intrinsic diode (0.7 V), and definitely quick, implements the freewheeling phase with less losses, and relieves the MOS during this phase. This Schottky diode becomes useless as soon as the MOS conducts (the conducting MOS voltage is much smaller than the one for the diode which blocks the diode). This diode must handle spikes in current.

3.5.4. *Step-up converter*

3.5.4.1. *Layout*

Figure 3.33 is the typical schematic of a Boost converter. In this basic pattern, there are neither sensors (Rshunt and dividing bridge) nor the MOS management circuit which controls the MOS current and output voltage.

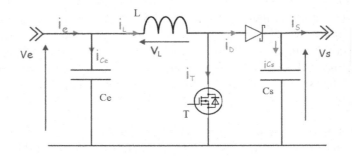

Figure 3.33. *Concept diagram of a boost*

Here we will also manage to minimize losses and the size of the devices. Control circuits working at a high frequency are used for this. We have, for example, used the LTC1872 which is a *Boost* controller working at a set frequency of 550 kHz.

Note that *T* and *f* are the switching period and frequency. α being the duty cycle, the transistor labelled T in Figure 3.33 is conducting for time interval [0; αT], and is open outside this interval.

3.5.4.2. *Choosing input capacitor*

For applications targeted in this work, we usually find a *Boost*-type converter located after the storage stage made up of supercapacitors. In this type of case, the input voltage of the *Boost* converter is equal to the ones in supercapacitors and varies very little in a switching cycle. However, we will place an input capacitor closest to the switching components that, as seen from the Boost input, will absorb the alternative component of the inductance current (Figure 3.34). The current provided by supercapacitors is therefore constant and equal to the average value of the inductance current. When inductance operates in continuous conduction mode, a calculation similar to the one made *supra* to scale output capacitor of the *Buck*, shows that rippling ΔV_{Ce} in the *Boost* input is independent of the duty cycle, and is expressed as:

$$\Delta V_{Ce} = \frac{\Delta I_L}{8.Ce.f} \qquad\qquad [3.11]$$

which gives:

$$Ce = \frac{\Delta I_L}{8.\Delta V_{Ce}.f} \qquad\qquad [3.12]$$

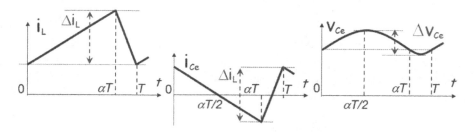

Figure 3.34. *Appearance of currents i_L, i_{Ce} and the voltage V_{Ce}*

3.5.4.3. *Choosing output capacitor*

The constraint is much higher on the output capacitor because the current in the diode is discontinuous. To determine the value of Cs, we assume that the value of this capacitance is enough to provide an output voltage that is practically constant. From 0 to αT, the diode is blocked and the capacitor alone provides the output current which is constant if the power needed is constant and if the output voltage varies slightly. We then get a constant current discharge of *Cs*, which gives the following expression for output voltage rippling:

$$\Delta V_s = \frac{<I_S>.\alpha}{Cs.f} \qquad\qquad [3.13]$$

where:

$$Cs = \frac{<I_s>.\alpha}{\Delta V_s.f} \qquad\qquad [3.14]$$

Rippling ΔV_S is set by the load specification and the average output current is given by the following relationship:

$$P_s = V_s.<I_s> \qquad\qquad [3.15]$$

With the switching frequency f having been chosen, only the duty cycle is left to be determined to set C_S. In continuous conduction, α is given by the relationship linking output voltage and input voltage:

$$V_s = \frac{1}{1-\alpha} V_e \qquad\qquad [3.16]$$

In the device's operational range, the voltage Ve varies as a function of the charge status in supercapacitors. One must then take a look at how the duty cycle varies to select Cs and guarantee for all situations a rippling of the output voltage that is in agreement with load specifications.

3.5.4.4. Choosing inductance

To limit the size of inductance, it is always good to work at the continuous/discontinuous conduction limit; rippling of the current in inductance is then equal to the current's maximum value and must remain less than or equal to the current's maximum value allowable by inductance without saturation.

The inductance value is classically given [SDR 06] by:

$$L = \frac{V_e.\alpha}{\Delta I_L.f} \qquad\qquad [3.17]$$

The power P_S required by the output load of the *Boost* is expressed by assuming a yield η for the converter in the form:

$$P_s = \eta.V_e.<I_e> = \eta.V_e.\frac{\Delta I_L}{2} \qquad\qquad [3.18]$$

This equation is used to determine the current's rippling needed to provide the power required by the load. Once the duty cycle α is determined by input and output voltages ratio (equation [3.16]), equation [3.17] allows for calculating the minimum inductance value needed.

3.5.4.5. Choice in MOS and in the Schottky diode

The Power MOS and the Schottky diode must be able to bear the same spike in current. When the diode is "ON", the MOS is subjected to an output voltage that it must bear; it will be the same for the diode when MOS conducts. As for *Buck* components, a safety margin is put on the maximum current and on the inverse voltage.

The greater the step-up of the *Boost*, the larger the duty cycle and the thermal constraint therefore increases on the MOS. It will be carefully chosen with a very small $R_{DS(on)}$. A loss analysis [PET 09] is needed to estimate lost power and thermally validate the choices made for encapsulation housing for devices.

3.5.5. Supplying power to converter electronics

DC/DC converters generally utilized for the energy converter and the level adapter (Figure 3.1) need a regulated electric power supply, whether it is for MOS drivers, supplying power to sensors or for control loops (analog or digital).

The power needed for this power supply is generally small compared to power transferred by converters. According to voltage levels available at the level of the energy harvesting system (V_{EH}) – EH for the *Energy Harvester* – and the desired level of power supply (V_{SS}), this power supply could be several different types.

3.5.5.1. $V_{EH} > V_{SS}$

In the case where voltage levels in the energy harvesting system are at a level greater than the level desired for electronic control (5, 3.3 or 1.8 V, for example), it is interesting to use an integrated linear regulator for its compact design (for example, type TPS799xx by Texas Instruments). We will make sure to verify that it can support a maximum voltage supplied by the energy transducer and that it could supply the required power.

A linear regulator's yield is often poor compared to a switching regulator; in fact, losses in a linear regulator are proportional to the voltage difference between the input and the output and to the current output. If this voltage difference is too big, a switching regulator will be better adapted to minimize losses. Fully integrated components (type LTC 3631) with a high yield carry out this function. They require the addition of an inductance and two capacitors. Since inductance dimensions remain small if the power connected is low, choosing a switching regulator to supply electronic power does not harm the set's integration.

3.5.5.2. $V_{EH} < V_{SS}$

In the situation where voltage levels in the energy harvesting system are less than the levels required for control electronics, as long as these levels remain compatible with circuit startup thresholds, we will take a *Boost*-type step-up layout to provide the electronic voltage power supply. Bear in mind that once enough voltage for the startup is achieved (*Minimum Start-Up Voltage*), 0.7 V for the LTC3539, for example, the *Boost* remains functional for a much lower voltage (0.5 V for this component).

3.5.5.3. $V_{EH} < 0.7V$

We have already approached this situation in section 3.3.2 for charge pumps that are particularly well adapted for very low levels of power with low starting voltages (voltages less than 0.7 V). In the same section, we show that the energy harvesting system could get its power from the voltage provided by the charge pump, and then directly from its own output when it has reached an adequate level.

3.6. Safeguards

An autonomous embedded circuit must be able to survive in its environment by dealing with human errors (bad connections between stages during assembly), stage malfunctions that it is linked to, and must continue to operate even if part of its own functionalities are no longer guaranteed. We will provide in this section several types of worst-case scenarios and likely solutions.

3.6.1. *Input safeguards*

In the commercialization phase, the input of a board linked to a solar panel will use a connector with a coded pin, to avoid polarity reversal. In the test phase without any specific connector, reversal is often possible, however, we can anticipate this configuration and create easily implantable solutions.

The first solution to reduce polarity reversal is to use a diode assembled in series (Figure 3.35), which blocks itself in case of polarity reversal.

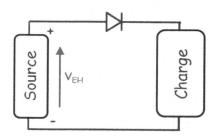

Figure 3.35. *Safeguard with a diode in series*

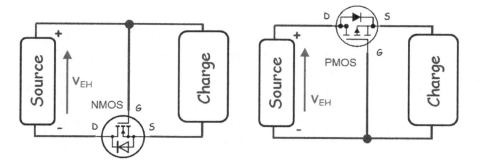

Figure 3.36. *Safeguard with an NMOS or a PMOS*

However, in normal operation, it lowers the useful voltage from 0.3 to 0.5 V and its power consumption that is often too high. We therefore prefer assemblies shown in Figure 3.36 where the diode is replaced with a MOS. Losses at the conducting state are much lower for the same current. The MOS is conductive in normal operation; it blocks in situations with polarity reversal. We find NMOS to be much easier, but its position compared to the negative line of the power supply is unfavorable and could lead to a preference for a PMOS placed on the positive line of the power supply.

Another solution to reduce polarity reversal is to use a Schottky diode with a low threshold, assembled in reverse. Normally blocked, it derives the current's entirety in a situation where there is polarity reversal (see Figure 3.37). It should then be scaled by current accordingly to take power from the harvesting system when the harvesting system sees a voltage V_F of about 0.3 V (voltage of the Schottky diode in a conducting state). In this figure, the device will not be operational but it will be protected.

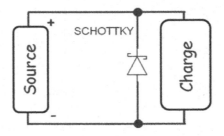

Figure 3.37. *Safeguards against polarity reversal from the source with a Schottky diode*

An elegant solution to avoid problems with input polarity reversal is achieved by using a full-wave rectifying diode bridge (PD2). Regardless of the connection, we will get a positive voltage at the system's terminals (see Figure 3.38). This solution is possible if the drop in voltage in the diodes ($2.V_F$) is not harmful to the system functioning properly.

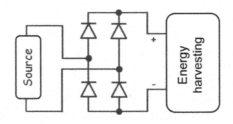

Figure 3.38. *Voltage harvesting with a full-wave rectifying bridge*

3.6.2. *Output safeguards*

Between the power pack of an autonomous embedded circuit and the load, it is good practice to place protective devices for overcurrents and voltage surges to protect either the power supply or the load according to the type of defect.

3.6.2.1. *Overcurrent*

The topology of *Boost* converters does not protect them against output short-circuits, so using a component dedicated to this functionality is therefore essential for maintaining system reliability [WIT 97].

In general, powering-up equipment leads to a big current in-rush during a short period. This current is induced by the load capacitor that is found at the input of equipment located downstream. The load's uptake power supply must then be scaled considering this in-rush current which could be a lot higher than the rated current. To avoid oversizing, this protective device is important as far as it is able to limit the in-rush current at startup (certain *Load Switches* have this property).

Finally, to avoid two cascaded stages starting up simultaneously, it is good to delay the startup of downstream stages. As such, with a *Boost* system supplying power to a load with energy from a supercapacitor storage stage, the load will not be powered until after setting up the downstream *Boost*, which is itself powered once the upstream *Boost* is started. We experienced that up-rush currents could lead to a system malfunction.

These different functionalities (short circuit, limiting currents, limiting up-rush currents and timed activation) are carried out by specialized devices like the LT1910 shown in Figure 3.39. This circuit drives an MOS that acts as a switch. A digital input can enable or disable the control circuit.

A few passive components around this electronic circuit allow the maximum authorized current to be set and to limit the up-rush current by utilizing the switch's progressive conducting setting (dV_{GS}/dt limited). When the authorized current is exceeded, the switch opens and protects the uptake stage. Periodically, automatic restart is initiated which allows a return to normal operation if the defect that led to an overcurrent has disappeared.

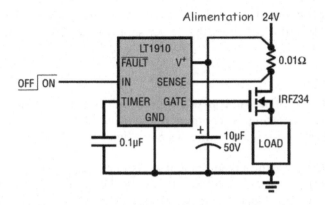

Figure 3.39. *LT1910 protected high side MOSFET driver (extract from its spec sheet)*

The set made up of a switch and its control could also be integrated into one component; an example with the ITS4141N by Infineon whose functional diagram is shown in Figure 3.40.

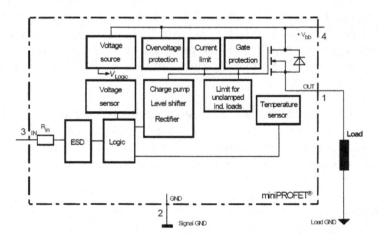

Figure 3.40. *Functional diagram of an ITS4141N "Smart High-Side Power Switch" (Infineon document)*

In the situation where we wish to be able to selectively drive several loads with independent overcurrent protection for each load, circuits such as the LT1161 are suitable.

Another option is to use a resettable fuse to protect against overcurrents. A resettable fuse (*polyfuse* or *polyswitch*) is a nonlinear thermistor that returns to a conducting state once the current has dropped, making it act as a timed circuit breaker and allowing the circuit to operate again. But these components have poor dynamic performance (outage times are a magnitude of a 10th of a second or more), and the maximum authorized current is too temperature dependent which is particularly limiting in aeronautical applications.

3.6.2.2. *Voltage surges*

To minimize the spread of overvoltages, the easiest solution is to use a surge arrestor made with a varistor or a Zener diode (Transil diode or TV for *Transient Voltage Suppressors* [VIT 07]) that carries out a clipping with a very low response time (a few hundreds of picoseconds for Transil diodes).

The proper Transil diode assembly is shown in Figure 3.41; we note the position of the fuse compared to the diode and the route configuration (very short connections between the diode and the power supply bus to limit inductive effects from the cables and tracks on the circuit board).

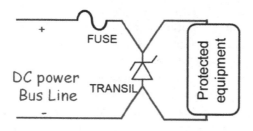

Figure 3.41. *Safeguard against overvoltage with a Transil diode*

Safeguards against overvoltages are also provided by components mentioned earlier, the ITS4141N, for example, protects against high voltages (>200 V) and ESD discharges (5 kV).

Dedicated components like type LT 4363 are also suitable according to the range of voltage and current involved in the application. As an example, Figure 3.42 shows the effect of the safeguard against an overvoltage by limiting the uptake overvoltage to an acceptable value set by the device characteristics.

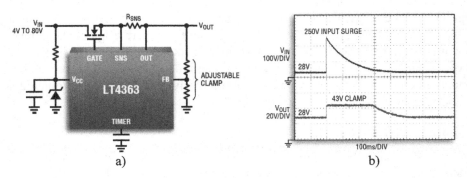

Figure 3.42. *a) Basic diagram and b) uptake in the downstream overvoltage*

3.7. Conclusion

In this chapter, we explained the advantages of supercapacitor-based storage as well as disadvantages induced by variability in voltage at the terminals of this device. For systems that are initially without any stored energy, we have also addressed solutions to put in place in order to successfully carry out an initial startup as soon as an adequate level of ambient energy is made available. More specifically, startup with a passive layout (often not powerful) was provided first. Then, we showed that more efficient energy harvesting system layouts exist. These layouts use DC/DC converters driven carefully in terms of the available resource (MPPT control). Finally, when the level of energy reaches a suitable threshold set by the application, the load can be powered and will remain so as long as the level of energy is above the set threshold. We have shown that activating the load could be implemented in different ways depending on the purpose. Finally, we provided possible technical solutions to protect the "supply" function and the elements that are connected to it.

The types of available energy resources, the intermittency or not of these resources and load consumption conditions (constant, variable) are extremely diverse. Additionally there is not a unique solution but several possible configurations for each considered project. Yet, energy management remains fundamental for such devices in order to ensure the best operation possible in accordance with operational conditions. Depending on the situation, these energy management functions (startup, switching over to another harvesting layout, load activation, active safeguards, etc.) could be decentralized and operate in an autonomous way, or in contrast, be managed in a centralized way using a microcontroller. In the framework of the overall design of a wireless system, the power supply function can also be connected to other parts of the system to inform, for example, the level of energy available in a way that enables eventually adjusting load consumption to optimize the operation period.

A one-size-fits-all solution does not exist, and the functions described in this chapter are therefore featured in different ways according to application scenarios. Chapter 4 gives three examples.

4

Build Achievements

4.1. Introduction

In this chapter, we will present three builds[1], the purpose of this is to clearly illustrate what was developed in previous chapters. The first two builds were installed in a passenger plane and were tested in-flight, while the third system was more of a start toward a solution for energy autonomy that is more innovative than previous ones.

With the two systems that were launched on the plane, design briefs required supplying power to *conventional* electronic housings, meaning they were partially optimized when it comes to energy consumption. The reader must not be surprised by the volume and weight of the systems designed for the management and harvesting of ambient energy. They meet power demands, depending on the situation where several hundred milliwatts to many watts are needed, and are therefore quite different to what academic research produces today, i.e. microsystems providing power that is lower by several tens. This also retroactively explains the choice in transduction methods that we made when writing the second chapter.

---

1 Our postgraduates Romain Monthéard and Paul Durand-Estèbe were key players in these builds.

Finally, it is worth mentioning that we will explain autonomous system design here and not results from measurement campaigns that they allowed us to make.

4.2. Autonomous power supply for external sensors in a flight testing campaign

4.2.1. *Flight testing background*

Flight testing (FT) is needed for development, evaluation and certification of new airplane models and their embedded systems. FT is also needed at manufacturing or maintenance output, and during client acceptance flights. They are carried out by specialized flight personnel.

FT most often involves setting up a test installation on the plane, often made up of sensors and devices registering and transmitting readings[2]. Hundreds of measurement points could be active during a flight[3], and the extra wiring could involve hundreds of kilometers of cables (see Chapter 1). Many test flights are required for certifying a new type of civil aircraft; for an Airbus A380 there had to be a little more than 2,000 h of flight and four test planes to get certification [LEL 13]. FT takes even longer and is more complex for a military aircraft. It is a heavy load for an aircraft manufacturer[4], who is always bound by scheduled delivery dates.

4.2.2. *Situations with external sensors*

In certain situations, sensors involved in FT are placed on the outside of the plane: acoustic noise measurement, readings on the propeller blades of a turboprop plane and aerodynamic pressure readings. In a situation involving wired instrumentation, once the issue with sensor placement is sorted out, a path must be made for the cables to go up into an entry point inside of the plane or directly inside the cabin (by replacing a rivet, for example, or often with a false window) to access a means for acquisition and registration. The

---

2 At least one part of these data is transmitted in real time to listening rooms on the ground.

3 Generating a substantial amount of data: in the range of 2 terabytes per day of flight testing with the Airbus A350, according to the *Journal Air & Cosmos* from 9 October 2015.

4 Around 40,000 € per hour of flight for testing the A380, which is 10 € per second of flight! [LEL 13].

question is especially critical as there are many sensors. The solution is to attach cables with a special glue, or with *aluminum foil tape*[5] for very temporary setups (see Figure 4.1).

In this context, we were approached to develop a solution for an autonomous power supply for aerodynamic sensors [DUR 16]. The design brief is given in the following section, but let us already mention the convergence of two very beneficial factors: on the one hand, the system (measurement and wireless transmission) to power is positioned on the outside of the plane, and on the other hand FT almost always takes place in daylight for flight safety considerations. A photovoltaic power supply is therefore naturally used as the ambient energy to recover.

Figure 4.1. *"Aluminum foil tape"-type external cabling on Airbus A400M. The area pictured is above the rear right lateral door called "parachutist door" that is visible in the bottom right of the photograph (Le Bourget 2011, photograph by authors)*

4.2.3. Principal elements of the design brief

Our project was justified by the airplane manufacturer's desire to substantially reduce the time span of testing campaigns by using sensors and measurement processing systems that are much easier to install, reposition and uninstall. More specifically, we had to design a power supply without a battery to power a pressure sensor device that transmits measurement data by

---

5 Knowing that sooner or later the adhesive will peel-off at one point, bail out and quickly breakdown.

radio. As already shown, we chose a photovoltaic harvester. The main elements in the design brief were the following:

– provide an average power of 2 W under 48 V;

– a temperature range of –50 to + 80°C;

– withstand a lowered pressure of 220 mbar;

– not to exceed 2–3 mm thickness for the solar panel, and 8 mm for the eletronics module for management and energy storage;

– modules should have a certain flexibility to be able to be attached to a variety of aerodynamic profiles;

– provide autonomy for a few tens of seconds in case of a loss of illumination (shadow on the wing during certain plane maneuvers when the sun is low on the horizon, or the impact of thick cloud cover).

Figure 4.3 shows the main structure of the entire design: photovoltaic panel, energy management electronics, sensors, electronics for signal processing and wireless transmission.

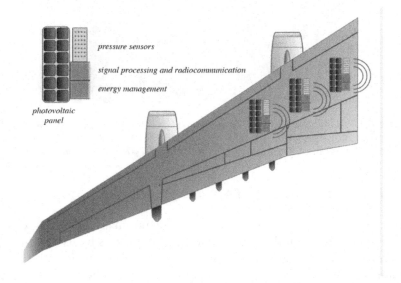

Figure 4.2. *Main layout of the different elements in the autonomous device to measure pressure. The electronic housings are streamlined, and could be a bit thicker than the photovoltaic panel because they are attached to the trailing edge of the wing, behind the sensors*

4.2.4. *Choosing technologies*

4.2.4.1. *Choosing major components*

The photovoltaic panel was chosen from among flexible technologies that are available (at the time of the study) giving the best efficiency possible, in order to limit the hold on the wing. However, as the device should operate as soon as the sun rises, we have tested several benchmarks under low light because solar panel efficiency is given by manufacturers under normalized conditions that correspond to strong sunlight, and an irradiance of 1 kW/m². At the end of testing, we selected the module SP 50L by SolbianFlex whose efficiency was close to or greater than 15% for irradiances varying from 57 to 822 W/m². This panel is 1.5 mm thick, weighs 790 g and measures 285 × 1,110 mm. It has 16 primary cells of silicon monocrystalline SunPower™, and, under standard light levels, it delivers 51 W[6] at its maximum power point (under 9 V and 5.7 A). Its minimum bend radius is 1 m.

With regard to electrostatic storage aimed at absorbing power spikes and to provide sufficient autonomy when illumination is lost, we tested several types of ultra-capacitors, giving priority to energy density compared to power density, the autonomy constraint of the design brief prevailing. Benchmarks for ultra-capacitors that we considered were reduced to prismatic geometries, the only ones that are likely to comply with thickness constraints. From a temperature perspective, no manufacturer could guarantee operation below −40°C. So, we proceeded with a testing campaign of charges and discharges at a constant current, for a temperature range starting from −50 to +100°C, and for some tests, pressure at 220 mbar. An example of the results is given in Figure 4.3 for the model selected, namely, the PC10HT by Maxwell Technologies.

This ultra-capacitor has a capacitance of 10 F and an ESR of 180 mΩ at room temperature. Its maximum operating voltage is 2.2 V[7]. It is guaranteed for a temperature range between −40 and +85°C. Its power and energy densities are 510 W/kg and 3.96 kJ/kg, respectively. Its mass is 6 g and its thickness is 4.8 mm with dimensions of 29.6 mm × 23.6 mm. We will use

---

6 For our application, the panel is significantly oversized for high irradiances.

7 Compared with the standard version, the HT version (High Temperature) of the PC10 is qualified for up to 85°C instead of 70°C, at the cost of a lowered working voltage, being 2.2 V instead of 2.5 V, a "conventional" value for an ultra-capacitor.

four of them[8] in a 2S2P combination, having a total capacitance of 10 F and a maximum working voltage of 4.4 V.

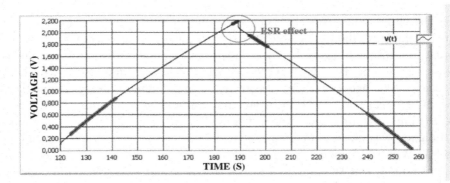

Figure 4.3. *Increase in voltage at 25°C at the terminals of a PC10HT ultra-capacitor during a charge and discharge at a constant current of 300 mA. Slopes in zones highlighted in bold let us get the value for capacitance, and the height of the discontinuity when the slope changes let us calculate the value for the equivalent series resistance (normally called ESR)*

At very low temperatures, even if the series resistance of the PC10HT rises, the ultra-capacitor continues capacitive behavior (charge and discharge profiles remain linear and the slope varies slightly) and therefore remains capable of storing energy. It does not suffer from a memory effect after high temperature cycles, and finally it does not deform under the effects of lowered pressure.

4.2.4.2. *Defining the electronic structure*

In the objective of operating at low irradiance (start and end of the day, overcast and flight in clouds), the energy management circuit must provide a good conversion efficiency at low power, while being able to operate at high irradiance, with a photovoltaic panel that is oversized in this case because calculating the panel surface is done so that the system operates at low irradiance. Measuring current combined with a protection circuit at the panel's interface is therefore needed.

---

8 This number is not the result of a scholarly scaling, but the outcome of a situational constraint in the fairing we had to utilize for the electronic system.

However, the energy management system must be capable of turning on in an autonomous and reliable way, with ultra-capacitors initially completely discharged (see methods proposed for this in Chapter 3).

Finally, for its build, it can only have components that are thin: with the issue with ultra-capacitors being sorted out, this point mainly affects the range of inductance values and therefore the values for regulator cutoff frequencies indirectly.

The overall structure is shown in Figure 4.4; it is an adaptation of generic structures for energy recovery and management circuits already given.

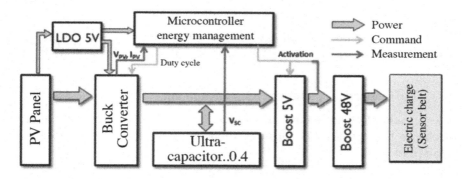

Figure 4.4. *Overall structure of a photovoltaic energy management circuit. Initially conceived to deliver 5 V, a late change in the design brief revealed this voltage to be 48 V shortly before flight testing, where there is a cascade of two Boosts at output*

Essential properties of this structure are as follows:

– a linear regulator (*Low Drop Out*, LDO) with low power supplies power to management circuits (current reader, MOS drivers and microcontroller);

– a *Buck* converter controls energy transfer to ultra-capacitors according to different modes (maximum power point tracking (MPPT) but limits the output voltage to 4.4 V to protect ultra-capacitors, current limitation is provided by the photovoltaic panel);

– two *Boost* converters in a sequence deliver a regulated output voltage of 48 V (non-optimal solution with two stages, led by a late change in the design brief);

– a microcontroller manages different operation modes.

We will now provide some details about key modules in this structure.

The DC/DC step-down converter (*Buck*) is a synchronous converter working in a continuous conduction mode. Its structure is given in Figure 4.5. Its switching frequency is 100 kHz and its major components are as follows:

– a transistor (Q1) MOSFET-type N acting as a switch;

– a second control switch (Q2) MOSFET N acting as a "conventional" freewheeling diode, a second control switch that is essential in a synchronous layout;

– 33 µH inductance, a little less than 3 mm thick;

– a Schottky antireturn diode intended to prevent ultra-capacitor discharge through the Q1 substrate diode in case irradiance is transitorily weak (darkness).

The converter is directly connected to:

– smoothing capacitors;

– two resistance bridges intended for the microcontroller to measure the voltage at the terminals of the photovoltaic panel (V_{PV}) and at the terminals of ultra-capacitors (V_{SC});

– a protection diode intended to protect the electronic system from polarity reversal when plugging in the photovoltaic panel;

– a controller by MOSFET (Microchip[Inc] MCP14628) making sure that there is synchronous commutation of Q1 and Q2 from the PWM (*Pulse Width Modulation*) signal generated by the microcontroller.

A measurement of the photovoltaic panel current flow, intended to protect the entire system against over currents, is provided by the circuit given in Figure 4.6. It links an amplifier (Texas Instruments INA 198) to a

100 mΩ shunt. Low-pass filters block the transmission of interference coming from converters.

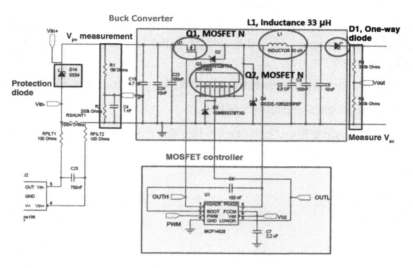

Figure 4.5. *Layout of a DC/DC step-down "Buck". The circuit appearing partially on the bottom left of the image is a current measurement circuit shown in Figure 4.6*

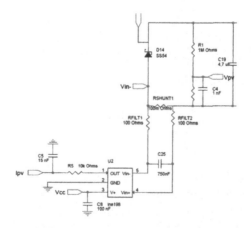

Figure 4.6. *Current/voltage measurement circuit. Connection to the panel is made at the protection diode level at the top left. The voltage measurement provided by the panel is carried out by the dividing bridge in parallel on the diode. Current measurement is led by the intermediary of the shunt in series with the diode*

The microcontroller and its immediate surroundings are given in Figure 4.7. It is the low consumption microcontroller PIC18LF1220 by Microchip[Inc]. Its functions are as follows:

– it executes the MPPT algorithm using the said method of *fraction of* V_{OC}[9] by regularly measuring the voltage at the terminals of the photovoltaic panel in an open circuit;

– it steers the circuit generating the PWM signal from the *Buck* converter;

– it controls ultra-capacitor loading;

– it activates the *Boost* exit stage when there is sufficient stored energy.

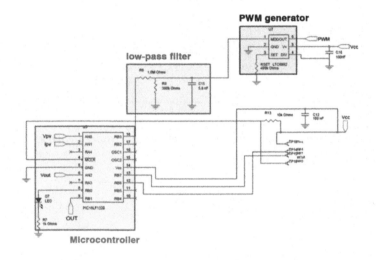

Figure 4.7. *The microcontroller and its immediate surroundings. The PWM LTC6992 generator is on top*

The PWM signal manages the commutation of MOSFET in the *Buck* converter and sets the commutation frequency to 100 kHz. It is generated by the integrated circuit by Linear Technology LTC6992 in Figure 4.7, whose frequency (100 kHz) is set with a simple external resistance. The size of impulses (duty cycle) is determined by a continuous signal obtained after

---

9 The most effective MPPT method at low irradiance.

low-pass filtration of a first PWM signal at low frequency coming from the microcontroller. The reason for this signal cascading is as follows.

We have set the frequency of the microcontroller's internal clock to 8 MHz to limit circuit consumption[10]. The generation mode, inside the microcontroller, of the signal setting impulse magnitudes will limit the number of discreet duty cycle values to 80 (if the signal frequency was 100 kHz), which appeared to be insufficient for the precision that is sought after. To work around this difficulty, the microcontroller directly generates a first PWM signal at 10 kHz only, but then has 800 discreet duty cycle values; this signal is then filtered, then sent to the LTC 6992 generator that itself generates a final PWM signal at 100 kHz. The combination represents a frequency transposition of the PWM signal. The filtering increases the time constant of the combination, which is not detrimental to our application[11].

Integrated control circuits supply power under 5 V using an LDO (Figures 4.4 and 4.8). This is a linear regulator by Texas Instruments TPS71550 connected directly to the photovoltaic panel. Since control components are not powered through ultra-capacitors, the circuit could startup when storage is empty. Voltage at the terminals of the panel has to be at least 6 V (voltage achieved with an irradiance of around 10 W/m²) for the Vcc power supply signal to reach 5 V.

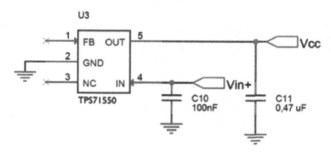

Figure 4.8. *Linear regulator supplying power to control integrated circuits through the V$_{CC}$ pin. The Vin pin is linked to the photovoltaic panel at the diode level that protects against polarity reversals*

10 Above this value, an external oscillator is absolutely required, which is unsuitable for consumption.

11 One must recognize that it is less important to optimize consumption in integrated circuits, when the amount of power delivered by the photovoltaic panel is high.

The output regulator in our application involves two very different stages, one delivering 5 V and the other delivering 48 V. This unbalanced structure with two stages is the result of a late modification of the power supply voltage for sensors used during flight testing. The output layout of our energy management system is the one described in Figure 4.9. The voltage (variable, but always maintained below 4.4 V) at the terminals of ultra-capacitors is regulated to 5 V by a step-up synchronous converter by Linear Technology LTC 3539 operating at a cutoff frequency of 1 MHz. The minimum voltage at input is 700 mV to minimize unused residual energy when ultra-capacitors are discharged. We have selected this converter for its high efficiency (function of delivered current), that is never less than 70%, and greater than 90% in our power range. It has a shutdown input logic $\overline{SHDN}$ (*shutdown*, active low) that we connected to two diodes carrying out an OR logic function. For the regulator to be active, at least one of the two anodes of the two diodes must be set to high level (HI). This is how the two following functions are carried out.

During startup, the microcontroller will not apply a HI to one of the diodes unless the load voltage (and hence stored energy) in ultra-capacitors is enough to support the transitory startup of output stages and supplying power to the load. This method is vital for a "clean and neat" regulator start-up (especially at low irradiance) and to avoid problems with "Under Voltage Lock Out" described in Chapter 3. The second stage, as we will see much later, will only start-up when the first stage applies 5 V at its input, avoiding a simultaneous startup of two stages.

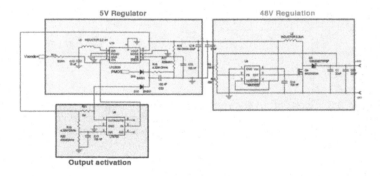

Figure 4.9. *The double output stage supplying voltage regulated at 48 V. The comparator at the bottom left manages the turn-on and shut-off at the first stage in conjunction with the two diodes forming the OR logic gate (see text)*

In the case where the photovoltaic panel is temporarily in the dark, the LDO and the microcontroller will cease to function very quickly. To avoid a premature stop of output converters, a comparator[12] by Linear Technology LT6700 continues to apply, through the second diode, an HI $\overline{SHDN}$ and keeps the circuit in operation as long as the regulator maintains a voltage of 5 V at output.

The conversion of 5 V toward 48 V is carried out with a Maxim MAX1523 regulator converter. Input $\overline{SHDN}$ of this integrated circuit is directly linked to the 5 V output of the preceding stage that controls activation. Two 22 μF output capacitors are tantalum capacitors that can support a 48 V voltage by maintaining a low volume.

4.2.4.3. Algorithmic aspects

The microcontroller's main task is to control the *Buck* converter according to readings taken: the photovoltaic panel's open circuit voltage, voltage and current flowing from the panel being loaded and ultra-capacitor voltage. The microcontroller will then adjust the duty cycle of the *Buck* converter according to one of the pursued objectives: MPPT, limitation of the input current and panel disconnection to measure its open circuit voltage. In addition, the microcontroller will activate the *Boost* output when capacitors are adequately charged (4.3 V).

From an input point of view, the system could then be in four different states:

– tracking the MPPT;

– limiting the current;

– at the end of loading;

– inactive, loading being complete.

The tracking of the maximum power point (MPPT) involves measuring the open circuit voltage V_{OC} every 10 s. The MPP is supposed to be at 81% of the V_{OC}, this value being predetermined experimentally, for the lowest irradiance and at room temperature. The Buck's duty cycle control is then adjusted to be set to this maximum power voltage. This MPPT algorithm

12 A comparator is used to generate a neat signal, leading to an immediate stop of the regulator.

based on a constant ratio to the measured open circuit voltage (sometimes referred to as the "open voltage" method) was preferred over other methods because of its robustness [MEE 10] (without realistically searching for a dynamic optimum, it cannot enter into an infinite cycle or find a local maximum) and ease of implementation. In addition, it needs to be efficient for only one context: low light and room temperature. Under other conditions, the power supplied by the panel is paradoxically too high to function at MPP.

If irradiance is too high, the current becomes too large to remain at MPP. It then centers the current limiting state. This state is activated when in MPP mode, the current reached its maximum value. The duty cycle is then adjusted so that the current always remains slightly below this maximum level. The current limiting stage is also active to manage the end of ultra-capacitor loading. At maximum current, the tension generated by ultra-capacitor ESR is significant. The objective is to progressively lower this current until either ultra-capacitors are fully charged, or there is a balance between the loading power and the power consumed by sensors connected to the system.

Loading is, on the other hand, completely stopped when ultra-capacitors reach their maximum voltage (V_{SC}=4.4 V). The system disconnects panels by setting the duty cycle to 0.

4.2.5. Builds and tests

The energy management system was carried out on a circuit board (PCB) 700 μm thick. Figure 4.10 shows a photograph. We then carried out ground-level integration tests, with a simulated load, and the photovoltaic panel placed under extremely variable sunny conditions[13]. The electronic device was then inserted into a fairing (Figure 4.11) intended to be fixed to the airplane's exterior.

---

13 The electronic system's efficiency alone had been assessed at around 67%.

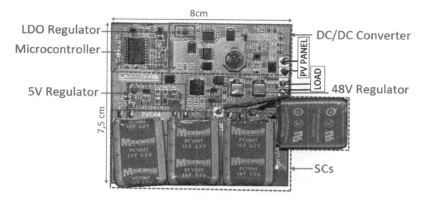

Figure 4.10. *Energy management system for photovoltaic applications. The fairing had dictated topology*

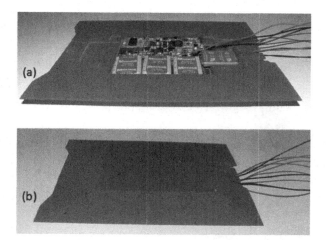

Figure 4.11. *Electronic card in its polymer fairing: a) viewed from below (plane's side wall) and; b) view from above (external side)*

The electronic device and the photovoltaic panel were then fixed on the wing of an Airbus A321 of the flight testing department at Airbus, at the same time as pressure sensors and measuring and telecommunications electronics that need to be supplied with power. Figure 4.12 shows the set up being assembled. Flight testing took place in the summer, but very early in the morning. Figure 4.13 shows a view of the decked out wing taken by a camera through one of the windows. We can see the low angled lights and darkness of the fin, indications of a precocious timetable and the need to be functional at low irradiance. Overall tests showed the relevance of the above design choices.

Figure 4.12. *Complete wireless system being assembled on the left wing of an Airbus A321: a) photovoltaic panel, b) energy management electronics and (c) pressure sensors and associated electronics (© Airbus)*

Figure 4.13. *System in-flight, on June 27, 7:08 GMT+1. The arrow shows the system's position (© Airbus)*

4.3. Autonomous power supply for age tracking sensors

4.3.1. *Introduction to structural health monitoring*

It is essential to regularly verify the health status of an aircraft and to then carry out the required maintenance operations in order to maintain a high level of flight safety in the world of civil aviation. Yet, maintenance operations represent, according to various sources, 10–25% of aircraft operating costs. This is particularly linked to the fact that when periodical inspection campaigns are carried out, they sometimes involve grounding planes for long periods to carry out essential disassembly and reassembly operations to access zones where, for example, corrosion or aging mechanical structures are likely to put the integrity of an airplane at risk. Yet the simple cost of grounding an airplane for an airline is $10,000 per hour[14], in addition to maintenance costs.

Within the scope of periodic inspections, mechanical safety margins are taken for some critical parts, guaranteeing that between two visits, regardless of what happens, no serious failure can arise. These margins increase the airplane's weight and the airplane's consumption of fuel indirectly.

The concept of *Structural Health Monitoring* (SHM)[15] has appeared over the past few years, which is the continuous monitoring of aging mechanical structures through the intermediary of permanently installed instrumentation. The concept is also recognized in marine engineering and civil engineering[16]. The goal is to move away from periodic inspections and to only launch maintenance operations when needed, by precisely identifying, in advance, elements that need to be changed: this is *preventative* maintenance. SHM sensors are very diverse, active or passive, and perform the measurement of various parameters such as mechanical stress, vibrations, acoustic emission, impedance readings, magnetic fields, etc. To our knowledge, the few aeronautical operational applications are today wired (and military, especially with drones).

4.3.2. Context of our study

We had to develop an energy harvesting system to supply power to sensors for SHM in the engine environment of an Airbus A380[17] (Figure 4.14) [DUR 16].

---

14 According to the *Journal Air & Cosmos* dated 9 October 2015.

15 Several related concepts coexist, such as the *Health and Usage Monitoring System* that aims to quantify the time remaining before a breakdown.

16 For example, refer to proceedings of the annual *International Workshop on Structural Health Monitoring conference*, Stanford, USA.

17 In the framework of the *Investissements d'Avenir* (Future Investments) program, and the EPICE platform (CORALIE project).

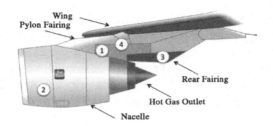

Figure 4.14. *The engine's immediate environment in a passenger airliner. (1) thermogenerator's location, (2) and (3) sensor locations, (4) location of the energy management electronics and signal processing as well as the local acquisition system*

This environment includes critical structural parts such as the nacelle, the pylon and its fairing. The goal of this project is to arrange SHM sensors in the nacelle and the fairing behind the pylon (*Aft Pylon Fairing*, APF), with wire-based sensors connected to a local acquisition system and then transmitting data to a concentrator. A module to harvest ambient energy and the electronics associated with it aim to make the local acquisition system autonomous.

The presence of permanent thermal gradients, found as soon as the engine is started, may encourage choosing thermogeneration as the source of electric energy. For instance in APF gradients, very large temperature spikes are achieved: one of the functions of an APF is rightfully to protect the upper parts of the pylon, as well as the wing, from rising hot air[18]. However, these temperature levels seem to be too high; moreover, estimating gradients is made more complicated by a significant influence from radiation transfers. We therefore discarded this localization. Subjected to high thermal and acoustic constraints[19], the APF on the other hand was retained to deploy sensors.

A decision was then made to install the thermogenerator device in the front of the pylon, in a more suitable thermal environment, guaranteeing an adequate gradient (see *infra*). The approximate location of SHM system components is detailed in Figure 4.14. It is the opportunity to discover an unknown and surprising characteristic of wireless sensor networks in an

18 On the A380, the APF is a mechanical assembly of around 4 m long and a mass of 110 kg, which also participates in aerodynamically shaping the vortex at the engine's output.
19 See the example *Airworthiness Directive* by the EASA 2010-0105R2, dated 5 March 2014.

extreme environment: the wireless network is basically cabled locally, over potentially large distances (here several meters). Basically, sensors are placed in close proximity to mechanical parts that they must equip, parts that are by definition in aggressive environments (temperature-wise here) and incompatible with electronic devices. The ambient energy recovery system itself obeys another constraint: being placed along the path of an energy flux (a thermal gradient in our case). Finally, electronics cannot generally handle temperatures that are typically greater than 85°C.

For functional reasons, the energy harvesting system and sensors must therefore (at least in the application developed here) be placed in distant locations, since because of the environment, electronic components will themselves be localized elsewhere if temperature ranges do not match (see Figure 4.15). Of course, less constraining situations could exist nonetheless.

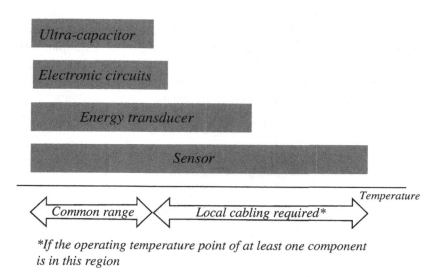

*If the operating temperature point of at least one component is in this region

Figure 4.15. *Various operating temperature ranges for different components of a sensor network node (relative positions are given as an illustration)*

It could be that, as in the application presented here, sensors, energy and electronic transducers are located in several different places and connected to one another by cable. This situation is given in Figure 4.16 which should be compared to *concept* Figure 1.1.

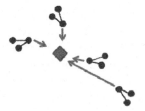

Figure 4.16. *Because of operational and environmental constraints, a sensor network called "wireless" could be cabled locally. This figure to be compared with Figure 1.1*

4.3.3. *Design brief*

The consumption profile of the acquisition system that our ambient energy harvesting module must make autonomous is a key element in the design brief. Its appearance[20] is given in Figure 4.17, knowing that during a flight only two transmissions take place with the data concentrator: one when engine starts, to initialize the system (for example, dating measurements), and the second at the end of the flight when the engine stops, to transmit important data coming out of the sensors. Note that the concentrator receives information from several acquisition systems.

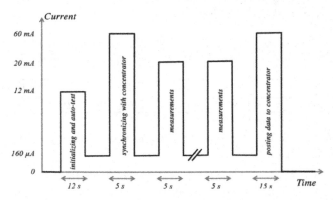

Figure 4.17. *Typical consumption profile for loading power generators. The load is supplied power under 5 V. The scales are not respected*

---

20 The acquisition system is not developed by the authors of this work, and it is deliberate that only combined data are provided here. This intended inaccuracy is not disadvantageous to the reader.

This profile allows us to verify that it is the data transmission that consumes the most energy, leading to 300 mW power spikes. Measurements matched with 100 mW spikes, but because of their low repetition, the average power consumed is only 1 mW. As described in Chapter 2, and especially as illustrated in Figure 2.4, we opted for a thermogeneration module supplying an average power greater than 1 mW coupled with storage from ultra-capacitors scaled to provide power demands of 100 and 300 mW.

The SHM system is not active when the engine has stopped; it is therefore not useful to scale storage for usage that is out of phase with the energy harvesting system.

Finally, in addition to *usual* compliance to the DO-160G standard (see Appendix), the particular sensitivity of zones labeled *1* and *4* in Figure 4.2 gave us particularly strict conditions in terms of fireproofing (1,100°C for 15 min); we will see in the following sections how we realized this during the design phase of our modules.

4.3.4. *Thermogeneration module*

The thermogeneration module (Figure 4.18) is installed in a cell in the pylon, identified to provide an in-flight thermal gradient (wall/inside air) that is adequate enough to generate the required average power, regardless of the flight conditions (temperature of outside air). The module is placed on a titanium wall 6.4 mm thick belonging to the primary pylon layout. The thermogenerator is in contact with this wall through the intermediary of a steel bearing plate set by a screw/nut system.

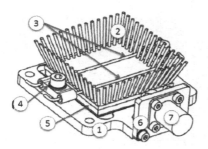

Figure 4.18. *Perspective view of the thermogeneration module (design by Xavier Dollat, LAAS-CNRS). (1) Plate, (2) exchanger (only one part of its shafts is shown), (3) flanges, (4) flange-holders, (5) thermogenerator, (6) connector-holder and (7) electric connector*

The thermogenerator is sandwiched between this plate (304 L stainless steel, melting point 1,454°C) and an exchanger (99.9% copper, melting point 1,084°C), and is placed under pressure by the intermediary of two flanges (pure annealed titanium, melting point 1,668°C). The electric output of the thermogenerator is linked by crimped steel wires sheathed in Teflon (melting point 327°C) to a *high temperature* electric connector certified for aeronautics. This connector is set in the plate by an intermediary made from a high temperature polymer support (Duratron PBI®, thermal and electric insulator, with an operation temperature of 310°C, possible transitional temperature of 500°C), this is to lower the temperature of the connector. This polymer is also the material used for the inserts in flange-holders to lower thermal bridges.

The assembly was designed in such a way that in the case of fire (the standard assumes that the wall of the pylon is brought to 1,100°C), as soon as the thermogenerator's semi-conducting elements (Bi_2Te_3, melting point 573°C) and materials made of polymers will have melted or burned, the module remains attached to the wall of the pylon, and that especially all of its metallic pieces as well as the thermogenerator's ceramic plates (Al_2O_3, melting point 2,050°C) remain interdependent. Digital simulations have shown that the copper exchanger will not melt under conditions imposed by the DO-160G standard.

The module shown in Figure 4.18 was successfully subjected to the vibrational spectrum predicted by the same standard (see Figure 4.19). The exchanger itself underwent a specific resistance to acoustic noise test (175 h at an acoustic intensity greater than 160 dB_A) without any detection of damage.

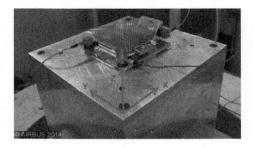

Figure 4.19. *Thermogeneration module on a vibrating pot during tests required by the DO-160G standard. Accelerometers were placed on the module for measurements (© Airbus)*

The module was installed on an Airbus A380 MSN01[21], in position 4 (right exterior engine), this position being equipped with a Rolls Royce Trent 900 engine. Flights were mainly organized within the scope of the *Flying Test Bench* program (in-flight test bed) to test (in position 2, left interior[22]) the new Rolls Royce engine with the new standard version of the A350 1000XWB.

Figure 4.20. *The four-engine Airbus A380 MSN04. It carries 22 tons of flight testing equipment (Farnborough 2012, photograph by authors)*

For the selected position for the thermogeneration module, the thermal gradient between the wall of the pylon and the air inside was evaluated at cruise level between 120 and 180°C, according to weather conditions and the flight level[23]. The wall temperature could itself exceed 250°C on the hottest days (*Maximum Hot Day*). Having made the option for a *conventional* thermogenerator layout (see Chapter 2), our choice of provider was found to be very restricted by the temperature performance that we were looking for (most models having an operational temperature well below 250°C), and was focused on a preseries model from the Japanese company *Kelk Komatsu* (see Figure 2.14) that we fortunately already maintain relations with[24].

---

21 MSN stands for *Manufacturing Serial Number*; the MSN01 is the airplane for the first flight of the A380, used for flight testing since the beginning of the program. It was named the *Jacques Rosay* in 2015, after the pilot who flew the first flight and passed away that same year.

22 In a *Flying Test Bench* configuration on a four-engine, it is always an interior position that is fed back to an engine being tested, to lower thrust imbalance in case of a breakdown [LEL 13].

23 The reader will excuse us again for not being able to be more precise for obvious trade secret reasons.

24 As we mentioned in Chapter 2, this illustrates the fact that choosing a thermogenerator could be linked to criteria other than its efficiency.

A first estimate of the power provided by this thermogenerator was done by using the model shown in section 2.4.2.3 (equations [2.7], [2.10]). The thermogenerator was then tested in a climatic chamber (collaboration with the company called *Neotim*). In Figure 4.21, the thermogenerator is shown in an interim assembly during these tests: it was placed in a horizontal position, which is not good for convection transfers, but this position matched the one that it would have in the pylon. In a worst-case scenario, i.e. for a 110°C gradient, the device provided 39 mW at maximum power point, power sufficient enough for our design brief, even considering different uncertainties in the models, in the conditions of this climate test and of the real flight environment.

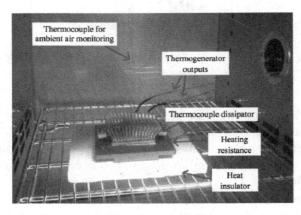

Figure 4.21. *Testing an interim version of the thermogenerator module in a climatic chamber. Tests were carried out at atmospheric pressure (photograph by authors)*

From a mechanical point of view, we proceeded to make a final optimization of the module, being aware of the final setting mode on the pylon, and searching for mass reduction. A photo of the final module is shown in Figure 4.22. We will see the differences compared to Figure 4.21 (where the plate is a copper beryllium alloy, with a melting point that is too low for fireproofing) and to Figures 4.19 and 4.21 (where the plate was cut on its inner face to make room for an attachment mode that was dropped in the final version).

Figure 4.22. *Photograph, top: three prototypes of the thermogeneration module showing progressive inclusion of weight, fireproofing and attachment constraints. Bottom: detailed view of the final version of the thermogeneration module. The total weight is 497 g. Considered as an energy generator, a female connector was used (photographs by authors)*

4.3.5. Electronic energy management system

Already shown functionally in its concept in Figure 2.3, we describe the layout and the design brief of the electronic energy management system. This system must:

– fulfill a form of impedance adjustment with the thermogenerator to maximize energy transfer;

– store enough energy to absorb consumption spikes exceeding immediately available power at the thermogenerator's output;

– provide a regulated voltage of 5 V;

– efficiently manage system turn-on (see Chapter 2).

Giving priority to robustness and reliability, at least because repairs during the flight testing campaign were difficult, we went ahead with the choices given below:

– using integrated circuits, even if efficiency is not as optimized as it could have been by building discreet components;

– choosing only one provider for active components (*Linear Technology*, specializing in embedded systems);

– protecting the output against short-circuits (loading failure or improper handling during final assembly in aircraft);

– common mode filtering at input, filtering aimed at protecting the system against the effects of lightning striking the aircraft.

The overall structure is shown in Figure 4.23.

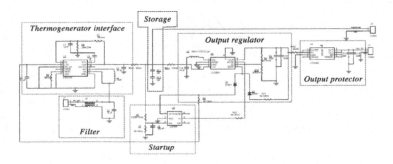

Figure 4.23. *Overall structure of the energy management system*

A common filtering mode aimed at protecting the system against conducted electromagnetic interferences was placed at the electronic card input (see details in Figure 4.24). These interferences would be harvested by unshielded parts of the thermogeneration module and caused by lightning striking the equipment, for example. It has a DC resistance of 80 m Ω and of 55 Ω at 1 GHz.

There is an LTC3105 integrated circuit at the filters output which is a DC/DC *Boost* regular (step-up voltage, inputs between 0.225 and 5 V, adjustable output from 1.5 to 5.25 V) guaranteeing three functions here (see Figure 4.24):

– it operates at a maximum power point (using a predefined setpoint, that is to say without active tracking) by controlling the input voltage Vin (setpoint adjusted by the external resistance R3). According to thermal data from the environment and the model of the thermogeneration module, we calculated a trade-off of value for this voltage (320 mV);

– it stops ultra-capacitors connected to the output from loading when the voltage Vout at terminals reaches 4.6 V (it is in fact the target value of the regulated output Vout, adjusted by bridge resistances R1 R2), which are protecting the ultra-capacitors (see *infra*);

– it provides, very early on, (as soon as Vin reaches 225 mV) a regulated auxiliary voltage (focused on polarizing logic circuits, maximum current of 6 mA) using a LDO regulator (*Low Drop Out*) adjustable with a resistance bridge (in our case, the LDO voltage is set at 2.2 V by grounding the FBLDO input – see Figure 4.24). The LDO voltage will be useful for us to provide power to the startup management circuit.

Storage is fulfilled by two BCAP0003 ultra-capacitors by Maxwell, with a unit capacitance of 3.3 F, maximum voltage of 2.3 V and an equivalent series resistance of 290 mΩ. These ultra-capacitors are connected in series, which lets us obtain a storage element with the value equivalent to 1.65 F and a maximum working voltage of 4.6 V, equal to the regulated output voltage of the converter that was just discussed. These components, while they store electricity in an electrostatic form, are affected by extreme temperatures. The benchmark is, for example, given for a –40 and + 85°C range, too narrow compared to what is required by the DO-160G standard (see Annex), which is between –55 and + 85°C. We therefore took the precaution to validate it experimentally, with testing, even if ultra-capacitor performance was even slightly deteriorated in a reversible way, allowing for us to operate at temperatures below –40°C. It is of note that the selected positioning in the airplane for the electronic system means that – excluding prolonged engine shutdown – it will never function at a negative temperature.

In the energy processing train, we then inserted the output regulator (Figure 4.25).

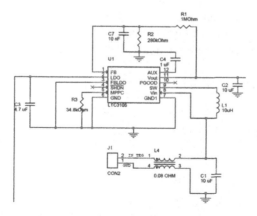

Figure 4.24. *Details of the input regulator and the common filter mode*

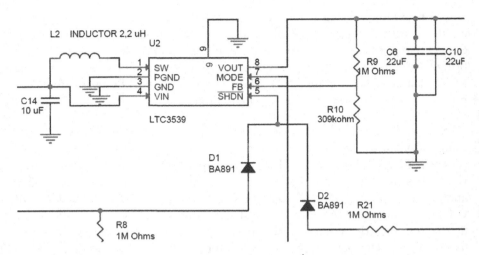

Figure 4.25. *Details of the output regulator and the turn-on logic*

Regulating the output voltage is fulfilled by an LTC3539 integrated circuit by Linear Technology which is a *boost* regulator (step-up), whose output voltage in our case is set at 5 V by bridge resistances R9 R10. It has a set cutoff frequency of 1 MHz, which is relatively high to reduce the value of L2 inductance. This integrated circuit was chosen for its high efficiency (more than 80% for the range of charging currents we are dealing with) and its capacity to operate with a low input voltage (typically 700 mV at startup, and up to 500 mV during operation). These two properties make the energy recovered by the TEG very profitable: good efficiency is required to minimize losses between storage and the load that requires power, and a low operating voltage makes use of the maximum energy stored in ultra-capacitors. The data below allows formula [2.3] to be applied in the current case, being:

$$E = \eta \; 1/2C \; [u_{max}^2 - u_{min}^2] = 0.8 \times 0.5 \times 3.3[4.6^2 - 0.5^2] = 27 \text{ J} \qquad [4.1]$$

The value that is obtained is enough in terms of the profile goal given in Figure 4.17, where, for example, the final transmission of data only consumes 4.5 J.

A logic input $\overline{SHDN}$ activates the output regulator by applying an LO logic and by avoiding an early startup (by supplying power to the load) known to be energy consuming, and potentially responsible for under voltage lock-out. More specifically, we will wait for a sufficient energy reserve before supplying power to the load. For this, a comparator (LT6700, found in the startup cartridge in Figure 4.23) generates an HI at the anode of the T1 diode (Figure 4.23) achieving along with diode D2 an OR circuit. Comparator switchover takes place as soon as voltage at ultra-capacitor terminals reaches 2.5 V (values set by the R6R7 bridge seen in Figure 4.23). The output regulator is therefore active and the load is powered[25].

In the hypothesis where the thermal gradient disappeared and where no more energy could be captured, the function OR would maintain the power supply to the active load as long as the regulator manages to maintain 5 V at its output, allowing us to take maximum advantage of the energy reserve in ultra-capacitors.

Finally, and while the regulator already has integrated safeguards against overcurrents, we added, in an interface with the load, a specific LTC4362 protection circuit that has an N channel MOSFET that acts as a switch between its input and output. The transistor's and measurement shunt's internal current resistances are low (40 and 32 mΩ, respectively) to minimize losses. If the current reaches 1.5 A, or if the input voltage reaches 5.8 V, the MOSFET opens and therefore disconnects loading. It switches back to a conductive state automatically at 130 ms.

Figure 4.26 gives the electronic card built according to the diagram in Figure 4.23. To fulfill the DO 160G standard, it was coated (with Electrolube DCA SCC3) in order to protect against fluids, and to prevent electric arcs (dielectric rigidity of the coating: 90 kV/mm). The card was integrated into a 316 L stainless steel housing (see Figure 4.26) to comply with the fireproofing standard (which a much lighter aluminum housing could not do). Aeronautical connectors (Souriau, series 2997 and 3649) are finally used in the connection between the thermogenerator and the load.

4.3.6. *Energy management system testing*

We will show the most important thing about electric testing here: the effects of temperature (operational test) and electromagnetic compatibility measurements (DO 160 G standard).

---

25 A similar layout was used in the first practical case dealt with in this chapter.

Figure 4.26. *The electronic energy management system in its housing. The total weight is 504 g[26] (build and photography by Microtec)*

4.3.6.1. *Temperature tests*

Figure 4.27 gives the experimental device aimed at testing our system functionality in terms of temperature. This device is made up of:

– the environmental test chamber (*ThermoStream* by Temptronic Corporation) subjecting the system in Figure 4.26 (without housing) to a selected temperature for testing;

– a power analyzer (Agilent N6705B) that plays a dual role: it simulates the thermogenerator (voltage and series resistance) and the load (profile goal in Figure 4.17).

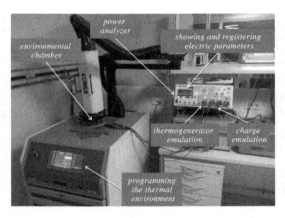

Figure 4.27. *Functional temperature test device for the electronic energy management system (photograph by authors)*

---

26 The attentive reader will make note of a surplus wire on the load side: its aim is to inform it about the state of the energy reserve in ultra-capacitors, and to eventually prevent launching procedures that could not be carried out to their term.

The decision to simulate the thermogenerator[27] avoids the presence of an extra thermal enclosure to create a thermal gradient (and not a simple environmental temperature). This absence does not hinder testing.

Tests were carried out for temperatures between –60 and +125°C[28], with intermediary points at –40, +23 (ambient) and +85°C. All components, apart from ultra-capacitors, are given by manufacturers for operating ranges between [–40, +85°C]. It is therefore likely that limitations come from ultra-capacitors.

In fact, without any surprise, we have realized that the system is no longer operational at –60°C; in fact the electrolyte used in ultra-capacitors is acetonitrile (solvent) with diluted salts tetramethylammonium/ tetrafluoroborate[29]. Yet the melting temperature for acetonitrile is –46°C[30]. At –60°C, this compound has undergone a phase change, and the ultra-capacitor no longer displays capacitive behavior. The system operates correctly at other temperatures, the crystallization of acetonitrile being a reversible phenomenon in terms of electric properties in ultra-capacitors.

The test at +125°C was brief (30 min maximum) and unique. It was only fulfilled as a particularity of the airplane manufacturer's design brief. Practically speaking, it is the temperature of the air in the environmental chamber that was brought up to +125°C, which is a lot higher than the maximum operating temperature (+85°C) given for semiconductors by their manufacturers, but perhaps without the full electrolyte reaching this

---

27 The thermogenerator here is simulated by a Thévenin generator with an open circuit voltage of 650 mV and a series resistance of 3.6 Ω.

28 These extreme temperatures are specified in the design brief of the aircraft manufacturer as being exceptional, and should be supported for 30 min without harming the outcome. The system does not have to be operational during this period.

29 These chemical compounds are toxic. Acetonitrile is also flammable and could enter your body by inhalation or by simple contact, breaking down into hydrocyanic acid (lethal) and thiocyanate.

30 The DO 160G cannot be complied with at this point, without having an impact on the airline's safety.

temperature. Hence, the test did not show any decline in system performance[31].

Figure 4.32 shows, infra, the standard operation of the complete system at room temperature, with the data logger as load to be powered.

4.3.6.2. *Electromagnetic compatibility tests*

The DO 160G standard requires several regulations to be respected in terms of disruptions that could be generated or subjected to by an electronic system, whether by conduction (via connecting cables) or by radiation. Proof of compliance must be carried out experimentally. The standard requires the use of unshielded cables for tests, since they will be that way in reality.

We then proceeded with tests, in an anechoic chamber, of the thermogenerator/energy management system/simulated load combination. The thermogenerator and the energy management system are connected here using a 4 m cable, while the energy management system and the load are connected with a 20 cm cable. These lengths being the ones predicted during installation on board the airplane. The load was made up of a resistance calculated to obtain the maximum load current (worst case). Interference was measured successively on two cables using a clamp ammeter (conducted interference), and through the intermediary of an antenna (radiated interference). Figure 4.28 shows the device used in the situation to measure conduction interference. These tests should verify that levels of radiant and conducted power were lower than the levels set for equipment in the engine's pylon.

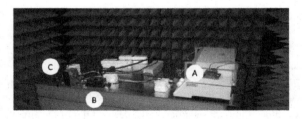

Figure 4.28. *Measuring conducted interference in the anechoic chamber. A: thermogenerator on heater, B: electronic energy management system, C: sensors to measure current (photography, Marise Bafleur)*

---

31 The ultra-capacitor dielectric probably did not have enough time to reach +125°C during this unique and short test.

Figure 4.29 shows the signal spectrum measured on the cable between the energy management system and the load. The template to follow is the one for the bottom-line labeled *power line*. Even in our context of DC power supply, the power signal is relatively noisy, while it conforms to the set template. Additionally, at 1 MHz a first line appears above the noise floor: it comes from the LT3539 output regulator's high current switching frequency (see Figure 4.13). Many lines are also found in the harmonic frequencies of the first. It is obvious that for much greater power supplies, the energy management system structure should be adapted to reduce conducting interference.

The cable placed between the thermogenerator and the energy management system did not emit any noticeable interference.

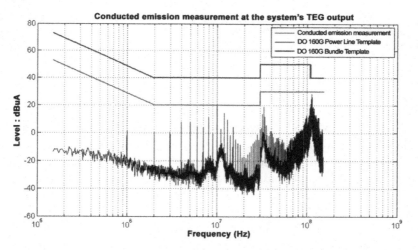

Figure 4.29. *Measuring conducted interference generated by the energy management system: signal experimental spectrum and template of the D0 160G standard (meaurement by Akexandre Boyer)*

The radiating signal was measured by a bilog-type antenna, oriented either vertically or horizontally toward the energy management system. Figure 4.30 gives the spectrum of this signal in a horizontal position. The template of the D0 160G standard is mostly observed here, as it was for the vertical component of the signal (not shown here).

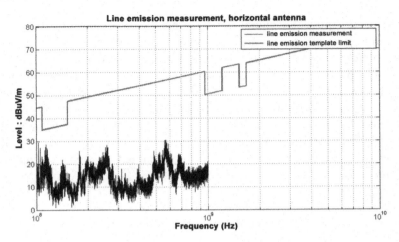

Figure 4.30. *Measuring interference from radiation (horizontal polarization in this example) generated by the energy management system: signal experimental spectrum and template of the D0 160G standard (meaurement by Akexandre Boyer)*

4.3.6.3. *Integration test*

With the objective to verify interface compatibility and good power switchover functionality[32] between supplying power to the data logger using an onboard network (initial state and stopped engines), or using the thermogenerator and the energy management system (once the engine is running and the thermal gradient established), these modules were connected electrically, while the thermogenerator was subjected to a thermal gradient representing the gradients expected in flight. The cables used had a layout (twisted wires and shielded cables) with lengths identical to those of cables expected to be used subsequently for assembling on the airplane (Figure 4.31).

To avoid untimely oscillations between the two energy sources, the data logger was programmed to switchover from the continuous 28 V power supply (identical to the one in the onboard network) with the 5 V power supplied by the energy management system (and inversely) through a hysteresis cycle, where thresholds were set for this test at 1.5 and 4 V (these thresholds involving voltage at the terminals of ultra-capacitors). In addition,

---

32 This switchover and this double power supply are presented here within the scope of a research and development project. A system that is truly autonomous in energy will definitely not be connected to an onboard network.

the data logger was configured for a workload (number of sensors and sampling frequency) representative of its real usage and energy consumption.

Figure 4.32 shows the results obtained. Initially discharged (worst case), ultra-capacitors charged up gradually, and when the voltage at their terminals reached 2.5 V the output regulator then delivered 5 V. However, it is not until the charging voltage reaches 4 V that supplying power to the data logger switches over from 28 V (airplane network) to 5 V.

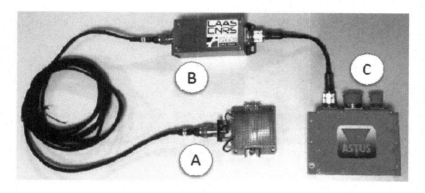

Figure 4.31. *A: integration test for the thermogenerator set, B: energy management system C: data logger connectors that are not connected here are used to collect data coming from sensors and to link them with the onboard network (photograph by authors)*

The thermal gradient was then voluntarily suppressed, and ultra-capacitors were gradually discharged; the data logger switched over again to a power supply of 28 V as soon as it crossed the 1.5 V threshold. During change-over switching between sources, there was neither loss of information nor untimely resetting to zero when operating the data logger. Beyond verifying functionality of the combination, it was noted that the "high" switchover threshold could be reduced in such a way to make supplying power by thermogeneration of the data logger active much sooner.

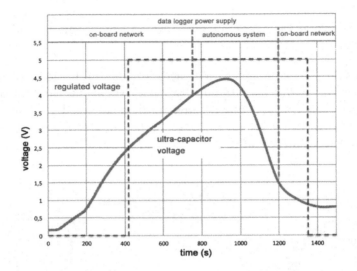

Figure 4.32. *Data logger power supply switchover test. The data logger switches in autonomous mode (regulated 5 V power supply) when ultra-capacitor charging voltage is between 1.5 and 4 V. Outside this interval, it is powered by the airplane's 28 V network*

4.3.7. *Airplane assembly and flight testing*

The set of equipment was then mounted on the Airbus MSN01[33], in the engine pylon of engine 4 (outside right). Figure 4.33 shows the thermogenerator before electric connection. The thermogenerator's plate was fixed to the titanium floor by a screw/nut system. To make thermal exchanges easier, the surface of the floor was cleaned beforehand and thermal grease was applied in a fine layer at the floor/plate interface. The energy management module and the data logger were placed higher in the pylon, in a zone that is less exposed to high temperatures.

At the time of writing, the energy harvesting system has completed "flight" for several months and currently meets design brief requirements.

---

33 Recall that flight testing took place within the scope of using the MSN01 as a flying test-bed for the new Rolls Royce Trent XWB-97 engine for the A350-1000.

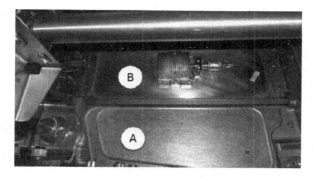

Figure 4.33. *The thermogenerator, once secured in the pylon. Note surface state differences between zones A and B (see text) (© Airbus)*

4.4. Aeroacoustic energy recovery

4.4.1. *Introduction*

In the broad context of application in the aeronautical domain, scenarios will involve situations where neither light energy nor thermal gradients will be a potential source (one that is available, reliable and lasting) of ambient energy. For reasons already mentioned, mechanical vibrations are also likely to be forgotten in many situations; airplane and parts manufacturers all have the goal, in each of their domains, to suppress or at least to reduce them. In certain situations, where the zone to be equipped is close to aerodynamic surfaces, it is possible to use an energy resource linked to relative wind (always present during the flight for fixed-wing aircrafts), at the cost of a very minor modification of these surfaces.

We now revisit the relevance in altering the airplane's external surface and violate its aerodynamics. Figure 1.3 shows that this is possible when the application justifies it.

To minimize aerodynamic interference, we tested a device that causes a transition from laminar to turbulent boundary layer[34], turbulent being a vibration mode that generates acoustic noise (called *aeroacoustics*) to be

---

34 Explicitly, we have renounced implementing wind turbines and any device that more or resembling them, blades or vibrating membranes, etc.

transformed into electric energy by a suitable transducer. Tests were carried out in a wind tunnel. The most important aspects will be discussed here, with details available in the references [MON 14a] and [MON 14b].

4.4.2. Concept

When air flows along a surface, it forms a boundary layer near the surface where the flow rate is reduced compared to the air velocity far from the surface U_∞. The lowered speed zone is called the *boundary layer*. A cavity located on the surface disrupts the boundary layer, leading to eddies or a vortex, and creates sustained oscillations. The impact of eddies on the cavity's downstream creates an acoustic wave that takes part in sustained oscillations. The situation is summarized in a simplified way in Figure 4.34.

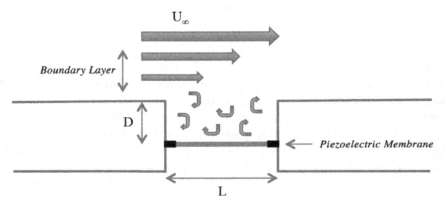

Figure 4.34. *Illustration of the generation of aeroacoustic noise. The cavity has a length L and a depth D. The piezoelectric membrane is used as a transducer*

4.4.3. Experimental results

Testing was not carried out in-flight, but in an ONERA wind tunnel[35]. This wind tunnel has a 50 mm square section on the side and allows us to introduce test vehicles which that are 30 mm long, a length that corresponds

---

35 It involved the wind tunnel and a B2A aeroacoustic measurement bed by ONERA able to operate between Mach 0.2 and Mach 0.6.

to the length of all the devices tested. Pressure measurement was carried out by using a microphone.

As a first step, we varied the geometric parameters L and D (see Figure 4.34), as well as the air flow velocity, to identify the most suitable configurations to create oscillations. This experimental approach was needed because modeling (even in two-dimensions) of the phenomenon is difficult. While taking into consideration the model by *Rossiter* [MON 14a, MON 14b], we can predict characteristic frequencies of oscillations, but not their amplitude.

We then identified the geometric configurations that produce the most energetic oscillations over the widest possible speed range. With these tests, we observed that:

– several geometries could achieve acoustic pressures at 150 dB[36], or mechanical power of about 100 mW/cm[2][37], a level of power that leads to a possible transduction of this mechanical energy into electrical energy, even with a low efficiency;

– the highest acoustic levels are found in the range of 2–3 kHz.

For the illustration that we provide *infra*, we used a cavity in which L = D = 27 mm: this cavity is very effective for speeds of Mach 0.4 and Mach 0.5[38], producing a level that is 152 dB, at frequencies of 2.26 and 2.34 kHz, respectively. To transform this mechanical energy into electricity, we placed a piezoelectric membrane (Murata 7BB-20-6) at the base of the cavity. The membrane's diameter is 14.4 mm. The electric signal spectrum harvested at the terminals of the membrane is given in Figure 4.35: the fundamental mode is predominant in terms of harmonics. The corresponding electric voltage is about 4 V peak, much greater than the threshold of a recovery diode.

---

36 Acoustic pressure in *decibels SPL* (Sound Pressure Level): Lp is defined by the equation $Lp=20log\,p/pref$ where p is the acoustic pressure and *pref* is the reference pressure (perception threshold of the human ear), equal to 20 µPa. The human hearing threshold is therefore 0 dB SPL. See also equation [2.11] and relevant comments.
37 At 20°C and at atmospheric pressure.
38 Remember that the Mach number is the ratio, in the given conditions, of the fluid's speed to the speed of sound in this same fluid.

With the aim to demonstrate the feasibility of the power supply to an autonomous system through aeroacoustic noise harvesting, we have built the following experimental device:

– the system already described *supra* (wind tunnel, cavity and piezoelectric membrane);

– an electronic energy management system (Figure 4.36);

– a *data logger* (Jennic JN5148) performing the temperature reading and transmitting the result by radio (protocol IEEE802.15.4, type ZigBee).

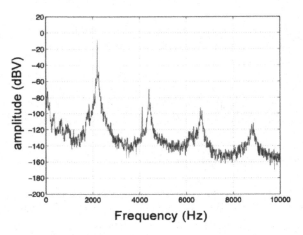

Figure 4.35. *Voltage spectrum in an open circuit at the terminals of the piezoelectric membrane (Mach 0.4)*

The electronic energy management system has a structure that is similar to the ones that we have already seen, while it is a bit more complex in that it is seeking a higher efficiency. It incorporates:

– an AC/DC converter recovering the voltage at the terminals of the membrane by using a synchronized switch harvesting on inductor (SSHI) scheme technique, which improves the conversion efficiency of the transducer;

– a DC/DC buck/boost converter adjusts (impedance matching) its input impedance compared to the output impedance of the AC/DC converter, and protects storage ultra-capacitors against over voltages;

− a storage stage made up of two ultra-capacitors in series (maximum 5 V), each with the capacitance of 0.5 F, linked to an active balancing circuit;

− a protection circuit for ultra-capacitors against overvoltage, OVLO (OverVoltage LockOut);

− a start-up management circuit UnderVoltage LockOut, whose function has already been described.

A photograph of this circuit is shown in Figure 4.37; inductors are the bulkiest components.

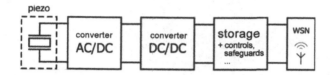

Figure 4.36. *Structure of the energy management system*

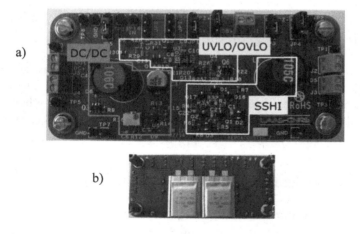

Figure 4.37. *a) Photograph of the energy management system;*
b) the view from below with two ultra-capacitors
(photograph by Romain Monthéard)

Figure 4.38 shows an overall view of the experimental device, where the only thing missing is the radio receiver connected to the portable computer located about 2 m away from the data logger. The experiment was carried out for a speed of Mach 0.4 and showed that the recovered aeroacoustic energy was sufficient to obtain the desired energy autonomy. A film of this experiment is available online[39].

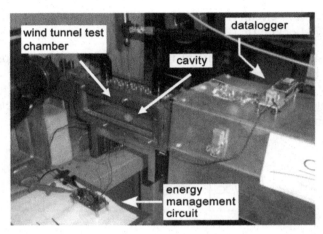

Figure 4.38. *Power supply device for a data logger by recovering aeroacoustic noise (photograph by Romain Monthéard)*

4.4.4. Conclusion: aeroacoustic noise harvesting

We have shown that harvesting acoustic noise created voluntarily was a conceivable solution to supplying power to circuits that are low energy consumers. It could be used for aerodynamic surfaces for which more conventional ambient energy sources are not available; the case of the landing gear could represent such an application.

4.5. General conclusion: build achievements provided

Examples given *supra* do not cover the entire range of conceivable applications. The reader would have understood that in terms of recoverable ambient energy and the load needing power, situations are extremely diverse. Nevertheless, functional similarities exist. It is the same for the development procedure requiring an identification and characterization step for available ambient energy, defining the consumption profile of the load

39 Available at: https://www.youtube.com/watch?v=1kkq-5g9D_g.

(profile goal) and finally, before every definitive choice, a feasibility study that eventually adapts this usage scenario when the energy demand exceeds what can be provided.

As demonstrated by the two major examples given in this chapter (flight testing and SHM): it is true that energy recovery for wireless systems could run into a short-term opening in the flight testing domain when an aircraft's safety is not an issue. Acquiring even greater maturity with the technologies (not only energy recovery techniques but also all elements of the network) is without a doubt required for operational uses of the wireless SHM without batteries.

Conclusion

As written in the Foreword, all of this work intends to give to the reader, who wants to keep up in an evolving domain of energetic autonomy, the keys that will open the right doors.

Many prospective studies predict significant market growth in these autonomous systems and related ambient energy harvesting systems: across all applicable sectors, anticipated financial amounts reach billions of euros by 2020 for the considered set of components (transducers, storage, management and energy conversion).

In considering only aeronautics, if the demands are great, short-term applications will bring on scenarios where flight safety for passenger airplanes is not involved. The reliability of the energy management system, as well as the system that is supplied with power, should provide a very high level.

However, flight testing during certification, for non-critical measurements, will probably be among the first applications involved, as well as the so-called "aircraft plant of the future". Large drones, including balloons and airships, could also be an applicable domain in the short-term.

At the same time, while in this work we did not approach the military domain, embedded systems using energy harvesting systems could find outlets in terms of arms, ammunition and decoys.

In this context, and in their own sector, the reader could – we hope – judge for themselves the relevance in continuing efforts in wireless sensor networks and capturing ambient energy.

Appendix

Summary of Certifications and Standards

Suggesting electronic solutions in the aeronautics domain involves adhering to many standards whose main objective is the reliability of the airline and the safety of its occupants. With regards to electronic embedded equipment, these must meet detailed criteria in the DO-160G standard of the RTCA[1], also known as the reference EUROCAE[2] ED-14G. This standard sets minimum performance requirements for equipment as a function of their nature, their use, the type of aircraft and where the equipment is located in it. We are not trying to provide details of this standard: the reader already being a practitioner must already know them, while the novice reader who must use his talents in the domain will become more familiar with them.

In this paragraph, we will use a few illustrations from the 20 or so technical sections of this standard, especially in the context of recovering and managing ambient energy. We develop certain aspects of qualifying equipment for real situations developed in Chapter 4.

Most importantly, environmental conditions to be followed are as follows[3]:

– an operational temperature that is between –55 and + 85°C;

---

1 *Radio Technical Commission for Aeronautics*, association mandated by the FAA for standardization.
2 *EURopean Organization for Civil Aviation Equipment*, European equivalent of the RTCA.
3 Obviously, this summary will not replace what is written in the standard.

– a minimum pressure of 116 hPa[4];

– a device's resistance, or at least the fasteners of a device on an aircraft, at an acceleration of 9 g for 3 s and an acceleration of 20 g for 20 ms;

– the device's resistance to vibrations whose spectrum (specified by the standard between 10 Hz and 2 kHz) depends on the location of equipment on the aircraft;

– the inability to ignite the surrounding atmosphere if the environment is explosive (areas with potential fuel leaks);

– resistance to humidity, fuel, hydraulic fluid, lubricants, solvents and cleaning liquids, salt, ice, mold, dust and sand;

– resistance to fire (does not create danger, for example by displacing or spreading flames, does not release toxic vapors), at 1,100°C for 5 min;

– good resistance to the direct and indirect effects of a lightning strike (specifically depending on the location of the device on the aircraft);

– low electromagnetic susceptibility, and limited electromagnetic emission (radiated and conducted);

– good resistance to electrostatic discharges.

Figure A.1. *An electromagnetic compatibility test of an electronic energy management system housing in an anechoic chamber, before being installed on an Airbus A380 (photography by Marise Bafleur, LAAS CNRS).*

4 Atmospheric pressure at sea level is 1,013 hPa. 116 hPa corresponds to an altitude of 15,000 m according to the atmosphere standardized by the OACI.

For specific cases (proximity to conducting fuels, absence of fire detectors and automatic extinguishers), constraints could be even greater. They could also be much lighter, for example for toxicity, when the mass of the device is reduced. Finally, within the framework of wireless autonomous energy systems, there is no issue with conducted electromagnetic coupling because these systems are not connected to the onboard network.

However, when searching for an adequate quantity of ambient energy, and for the considered applications, which often lead to consideration of extreme environments (for example, temperature and acoustic noise,), environmental conditions could be stricter requirements than those required by the standard. We illustrate this point in Chapter 4.

Concluding this short outline, we note that the tests involved in standardization require extremely varied techniques and skills that are rarely found in only one individual. This reinforces the statement made elsewhere in this book, that the design of an energy autonomous wireless sensor network for aeronautics is a multidisciplinary work which requires a holistic approach.

Bibliography

[ALB 08] ALBEROLA, PELEGRI J., LAJARA R. *et al.*, "Solar inexhaustible power source for wireless sensor node", *Instrumentation and Measurement Technology Conference*, Victoria, Canada, pp. 657–662, 2008.

[BAI 08] BAILLY N., DILHAC J-M., ESCRIBA C. *et al.*, "Energy scavenging based on thermal gradients: application to structural health monitoring of aircrafts", *POWERMEMS*, Tokyo, Japan, pp. 205–208, 2008.

[BAR 05] BARBU S., Conception et réalisation d'un système de métrologie RF pour les systèmes d'identification sans contact à 13,56 MHz, PhD Thesis, University of Marne-la-Vallée, 2005.

[BEC 08] BECKER T., KLUGE M., SCHALK J., "Power management for thermal energy harvesting in aircrafts", *IEEE SENSORS*, Lecce, Italy, pp. 681–684, 2008.

[BEL 12] BELLEVILLE M., CONDEMINE C., (eds.), *Micro et nanosystèmes autonomes en énergie*, Hermes-Lavoisier, Cachan, 2012.

[BIR 13] BIR U., "History of thermoelectricity", in JÄNSCH D. (ed.), *Thermoelectric Goes Automotive II*, Expert Verlag, Renningen, pp. 88–101, 2013.

[BOI 13a] BOISSEAU S., GASNIER P., GALLARDO M. *et al.*, "Self-starting power management circuits for piezoelectric and electret-based electrostatic mechanical energy harvesters", *Journal of Physics: Conference Series*, vol. 476, December 2013.

[BOI 13b] BOITIER V., DURAND-ESTÈBE P., MONTHÉARD R. *et al.*, "Under voltage lock-out design rules for proper start-up", *Journal of Physics: Conference Series*, vol. 476, December 2013.

[CAR 11] CARLI D., BRUNELLI D., BENINI L. *et al.*, "An effective multi-source energy harvester for low power applications", *Design, Automation & Test in Europe Conference & Exhibition*, Grenoble, France, pp. 1–6, 2011.

[CUL 14] MCCULLAGH J.J., GALCHEV T., PETERSON R.L. *et al.*, "Long-term testing of a vibration harvesting system for the structural health monitoring of bridges", *Sensors and Actuators A-Phys.*, vol. 217, pp. 139–150, 2014.

[DAM 97] DAMASCHKE J.M., "Design of a low-input-voltage converter for thermoelectric generator", *IEEE Transactions on Industry Applications*, vol. 33, no. 5, pp. 1203–1207, 1997.

[DAN 13] DAN A., ZHU M., TIWARI A., "Evaluation of piezoelectric material properties for a higher power output from energy harvesters with insight into material selection using a coupled piezoelectric-circuit finite element analyses", *IEEE Transactions on Ultrasonics Ferroelectrics and Frequency Control*, vol. 6, no. 12, pp. 2626–2633, 2013.

[DUR 16] DURAND-ESTÈBE P., Systèmes de récupération d'énergie pour l'alimentation de capteurs autonomes pour l'aéronautique, PhD Thesis, Institut National des Sciences Appliquées, Toulouse, 2016.

[FER 06] FERRIEUX J.P., FOREST F., *Alimentations à découpage, convertisseurs à résonance, principes, composants et modélisation*, Dunod, Paris, 2006.

[IOF 57] IOF A., *Semiconductor Thermoelements and Thermoelectric Cooling*, Infosearch Ltd, London, 1957.

[JAC 10] JACQUEMODA G., NOWAKA M., COLINET E. *et al.*, "Novel architecture and algorithm for remote interrogation of battery-free sensors", *Sensors and Actuators A-Physics*, vol. 160, pp. 125–131, 2010.

[JIA 05] JIANG X., POLASTRE J., CULLER D., "Perpetual environmentally powered sensor networks", *Fourth International Symposium on Information Processing in Sensor Networks*, Los Angeles, USA, pp. 463–468, 2005.

[KLE 14], KLEIN J., Synchronous buck MOSFET loss calculations with excel model, Application Note, Fairchild Semiconductors, available at: https://www.fairchildsemi.com/application-notes/AN/AN-6005.pdf, 2014.

[KOW 09 11] KOWAL J., AVAROGLU E., CHAMEKH F. *et al.*, "Detailed analysis of the self-discharge of supercapacitors", *Journal of Power Sources*, vol. 196, pp. 573–579, 2011.

[LEL 13] LELAIE C., *Les essais en vol de l'A380*, Le Cherche Midi, Paris, 2013.

[MEE 10] MEEKHUN D., Réalisation d'un système de conversion et de gestion de l'énergie d'un système photovoltaïque pour l'alimentation des réseaux de capteurs sans fil autonome pour l'application aéronautique, PhD Thesis, Institut National des Sciences Appliquées, Toulouse, 2010.

[MER 12] MERRETT G.V., WEDDELL A.S., "Supercapacitors leakage in energy harvesting sensor nodes: fact or fiction?", Ninth International Conference on Networked Sensing Systems, Antwerp, Belgium, pp. 1–5, 2012.

[MON 14a] MONTHÉARD R., Récupération d'énergie aéroacoustique et thermique pour capteurs sans fil embarqués sur avion, PhD Thesis, Institut National des Sciences Appliquées, Toulouse, 2014.

[MON 14b] MONTHÉARD R., AIRIAU C., BAFLEUR M. et al., "Powering a commercial datalogger by energy harvesting from generated aeroacoustic noise", Journal of Physics Conference Series, vol. 557, November 2014.

[PER 13] PERNAU H., JACQUOT A., TARANTIK K. et al., "Standardisation of metrology for thermoelectric materials", in JÄNSCH D. (ed.), Thermoelectric Goes Automotive II, Expert Verlag, Renningen, pp. 125–134, 2013.

[PET 09] PETIBON S., Nouvelles architectures distribuées de gestion et de conversion de l'énergie pour les applications photovoltaïques, PhD Thesis, University of Toulouse, 2009.

[PIC 08] PICHOT P., Optimize power consumption in portable electronics using integrated load switches, Portable Design, Texas Instrument, October–November 2008.

[RAV 11] RAVINDRAN S., HUESGEN T., KROENER M. et al., "A self-sustaining micro thermomechanice-pyroelectric generator", Applied Physics Letters, vol. 99, p. 104102, 2011.

[SAM 11] SAMSON D., KLUGE M., BECKER T. et al., "Wireless sensor node powered by aircraft specific thermoelectric energy harvesting", Sensors and Actuators A-Physics, vol. 172, no. 1, pp. 240–244, 2011.

[SIO 12] SION C., GODTS P., ZIOUCHE K. et al., "Unpacked infrared thermoelectric microsensor realized on suspended membrane by silicon technology", Sensors and Actuators A-Physics, vol. 175, pp. 78–86, 2012.

[SPI 08] SPIES P., POLLAK M., ROHMER G., Power Management for Energy Harvesting Applications, available at: http://www.researchgate.net/publication/228693238_Power_Management_for_Energy_Harvesting_Applications, 2008.

[SPI 15] SPIES P., MATEU L., POLLAK M., Handbook of Energy Harvesting Power Supplies and Applications, Pan Stanford Publishing, Singapore, 2015.

[VAN 15] VANHECKE C., ASSOUERE L., WANG A. *et al.*, "Multisource and battery-free energy harvesting architecture for aeronautics applications", *IEEE Transactions on Power Electronics*, vol. 30, no. 6, pp. 3215–3227, 2015.

[VIS 07] VISHAY INTER TECHNOLOGY, Failure modes and fusing of TVS devices, Application Note, Vishay General Semiconductors, document number 8440, available at: http://www.vishay.com/docs/88440/failurem.pdf, 2007.

[WED 11] WEDDELL A.S., MERRETT G.V., KAZMIERSKI T.J. *et al.*, "Accurate supercapacitor modeling for energy harvesting wireless sensor nodes", *IEEE Transactions on Circuit and Systems-II: Express Briefs*, vol. 58, no. 12, pp. 911–915, 2011.

[WIT 97] WITT J., Short-circuit protection for boost regulators, Design Note 154, Linear Technology, 1997.

Index

Printed in the United States
By Bookmasters